Renata Figueiredo Anomal

The evolution of the somato-motor system in mammals

Renata Figueiredo Anomal

The evolution of the somato-motor system in mammals

The sensory-motor amalgam of the opossum Didelphis aurita

ScienciaScripts

Imprint

Any brand names and product names mentioned in this book are subject to trademark, brand or patent protection and are trademarks or registered trademarks of their respective holders. The use of brand names, product names, common names, trade names, product descriptions etc. even without a particular marking in this work is in no way to be construed to mean that such names may be regarded as unrestricted in respect of trademark and brand protection legislation and could thus be used by anyone.

Cover image: www.ingimage.com

This book is a translation from the original published under ISBN 978-613-9-63356-2.

Publisher:
Sciencia Scripts
is a trademark of
Dodo Books Indian Ocean Ltd. and OmniScriptum S.R.L publishing group

120 High Road, East Finchley, London, N2 9ED, United Kingdom
Str. Armeneasca 28/1, office 1, Chisinau MD-2012, Republic of Moldova, Europe
Printed at: see last page
ISBN: 978-620-7-65875-6

DEDICATORY

"To Bianca and my grandparents, Aloir Rodrigues and Zenetti A no mal."

ACKNOWLEDGEMENTS

My thanks go initially to the primary cause of all things and the opportunity that life has allowed me to develop this work in such special circumstances.

I would like to thank each and every one of the experiences I have had during my time here in the Neurobiology II laboratory, as they have all been fundamental to my intellectual and internal growth.

I would like to thank all the people who have been part of this period. Those who have gone on to other places, those who are still part of this group, those who have accompanied me closely and those who have contributed more distantly.

In particular, I would like to mention my family for the support they have given me over the last five years.

I'd like to thank Bia, my greatest love, who I'm sure was fundamental to me completing my work.

I can't forget my friends, Nanda and Brian, who practically did the PhD with me, listening to me in those "Murf's Law" moments. I would also like to thank those friends who continue to be my friends, even with so many articles to read and experiments to do.

I would like to thank those who opened the doors for me to be here at the Institute today and helped me with my training, such as Vanessa Rocha, Professor Eliane Volchan and Professor João Franca.

I would like to thank my students Jorge Arigoni, Fabiana Santana and Fernanda Muniz for their help. May they have taken a bit of scientific thinking into their lives.

I would like to thank my fellow travellers, with whom I have shared good times, such as Eliã Botelho, Carlomagno Bahia, Cristiane Ashidamini, Ghislain Saunier and Anaelli Campos.

Finally, I would like to thank everyone who was part of our team, such as Luiz Bemardino, Paulo and Gervásio, because their contributions were very important to the development of this work; as well as all the members of other laboratories who helped me with my experiments, such as Professors Mário Fiorani and Jean Christopher Houzel.

May all the other people who were with us during this period of my life feel equally involved with these heartfelt thanks.

SUMMARY

In the present work we studied the pattern of thalamic and cortical projections to the parietal cortex of the opossum *Didelphis aurita* through injections of retrograde fluorescent tracers in the primary somesthetic (S1) and caudal (SC) areas. In order to characterise the rostral and caudal borders of S1, the electrophysiological and architectural characteristics of S1 and SC were also described in the anterior parietal cortex of this species Retrograde cell marking in the cortex and thalamus showed that S1 and SC have distinct connection patterns. Area S1 has intrinsic connections and receives projections from other somesthetic cortical areas, such as SR and SC. Projections from the secondary somatosensory (S2) and ventral parietal (PV) areas were rare. On the other hand, area SC receives cortical projections from the somesthetic, visual, auditory and frontal cortex areas, suggesting multimodal processing in this region. The thalamic connections to area S1 arise from the ventral basal thalamic "somesthetic" nucleus (VB), and the ventral lateral (VL), ventral anterior (VA) and ventral medial (VM) thalamic "motor" nuclei. Area SC also receives projections from VB, VL and VA, as well as thalamic groups that do not project to S1 (posterior group). The motor thalamic projections to S1 and SC suggest that cerebellar information is processed in both areas, without necessarily characterising them as motor cortex. This projection pattern observed in the opossum *Didelphis aurita* may represent the primodial organisation of the sensory-motor system in mammals, where areas involved in multimodal and motor processing, such as S1 and SC, could have originated the sensory-motor areas currently observed in most extant mammals.

SUMMARY

CHAPTER 1

INTRODUCTION

Not infrequently, we see athletes and musicians surpassing expectations by showing surprising sensory-motor skills that clearly differ from some of the stereotyped and simple behaviours of other mammals. The difference in sensory-motor abilities between mammal species is partly due to the morphology and physiology of each species. However, it is known that the organisation of the encephalon and neocortex play a special role in the execution and improvement of these skills (KAAS, 2008; KA AS, 2004).

The neocortex is a six-layered structure that covers the mammalian encephalon and is involved in a series of complex functions *(cf.* BRODMANN, 1909, p. 17). This structure is considered to be one of the most striking features of the evolution of the mammalian encephalon and was subdivided in humans and other mammals by Brodmann (1909) into functionally different "processing organs" called **cortical areas.** Each cortical area, in turn, connects with other areas and with subcortical structures, thus forming **cortical systems.** It is now known that some cortical areas of the somatosensory system have both somatosensory and motor functions (KAAS and STEPNIEWSKA, 2002). The way in which the sensory and motor systems interact characterises the nature of a species' sensory-motor abilities to such an extent that we can estimate its capabilities by knowing its cortical organisation (KAAS, 2008).

It is currently estimated that the human encephalon has more than 200 cortical areas, and that many of these areas are involved in sensory and motor processing (KAAS, 2008). In contrast, it is believed that ancestral species had a small encephalon in relation to their body weight, and apparently simpler sensory and motor systems, made up of few cortical areas (KRUBTIZER, 1995; KRUBITZER and KAAS, 2005; KAAS, 2008, 2004; CATANIA, 2005).

But how did the encephalon reach different levels of complexity over the course of evolution? Although this is a difficult question to answer completely, some answers can be obtained by studying the sensory and motor systems of extant mammals and analysing ancestral fossils.

1.1 Reconstructing the evolution of the neocortex in mammals

During a certain period, from 1880 to 1920, there was intense activity in palaeontology interested in the evolution of the brain (BUTLER and HODOS, 1996). Initially, the information obtained from fossils of ancient mammals, such as brain area and volume, was widely used for evolutionary studies (BUTLER and HODOS, 1996). However, considering only skull fossils, little information is provided to understand the internal organisation of the encephalon. This is because the brain is not fossilised and the information of interest is obtained indirectly through the marks left on the inner surface of the skull, with little or no information about its connections, architecture and cortical organisation (KAAS, 2008; KAAS, 2002; KAAS and COLLINS, 2001a).

The first attempts to reproduce a model of the ancestral mammalian encephalon from existing species used the **quasi-phylogenetic** criterion (CLARK, 1959; KAAS, 2002). According to this criterion, species are classified into levels of complexity, creating an evolutionary sequence according to the presence of certain encephalic characteristics that are considered primitive or not (SMITH, 1924; CLARK, 1959). However, taking into account that existing species do not have solely primitive or solely derived encephalic characteristics, this idea of progression in the encephalon in a single direction - from the simple to the more complex - does not correspond to reality (KAAS, 2002).

Cladistic analysis is the most reliable alternative for studying the evolution of the encephalon by comparing existing species (KAAS, 2002), as it classifies animals according to their phylogenetic relationships, which are established by the similarity of characteristics shared between species, grouping them into taxonomic groups. In this system of analysis, species are ordered in a cladogram or dendrogram and fit into branches (as in a tree), each representing a taxonomic group or a species, explaining the phylogenetic distances between the groups (KAAS, 2002).

Two elementary principles are used in cladistic analysis: the *comparison of outgroups* and the *principle of parsimony*. Outgroup comparisons are made between distinct species or groups, whether they are phylogenetically distant or close (BUTLER andHODOS, 1996).

In the principle of parsimony, the hypotheses generated by analysing groups or species are based on the smallest number of possible transformations between the species studied (BUTLER and HODOS, 1996). Thus, in order to investigate whether or not a given characteristic belonged to a supposed ancestral species, its frequency among the different taxonomic groups is analysed. Following the principle of parsimony, a characteristic is said to be homologous between species when it is observed in all or most of the species studied in a taxonomic group. This characteristic is then considered **plesiomorphic** because, according to the parsimony principle, it would have been maintained throughout evolution from the ancestral species (BUTLER and HODOS, 1996). It should be borne in mind that the simplest and most parsimonious hypothesis is likely to be the correct one, but this is not a guarantee of the best hypothesis.

Look at Figure 1 below. A certain "+" characteristic (the existence of a cortical area, for example) is present in species A, C and D, but not in B. The principle of parsimony suggests that the common ancestor of A and B would have possessed the "+" characteristic (because C and D have it) and that B would have lost it in its evolutionary process. But it is possible that this ancestor had this characteristic and that the subsequent species lost it before B appeared. Another possibility is that the ancestor of A and B didn't have this characteristic and A developed it independently. Note that these latter hypotheses suggest a greater number of changes than the first, which is considered the most parsimonious.

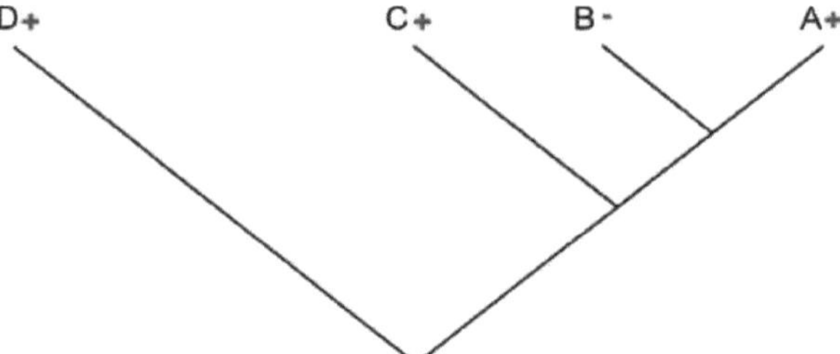

Figura 1.   Cladogram showing the distribution of a "+" characteristic among existing species.

As illustrated in Figure 2, the primary cortical areas involved in visual, auditory and somesthetic sensory processing are present in the main subgroups of mammals, such as marsupials, eutherians or placentals (rodents, primates and carnivores) and prototherians (omitorrinco and echidna) (KRUBITZER, 2007). It is therefore suggested, according to the principle of parsimony, that the presence of these areas in extant mammals is a plesiomorphic characteristic, maintained in the species from the common ancestral mammal.

The particular characteristics of certain species, which are not widely distributed in the cladogram, are supposed to have arisen independently (ELDREDGE and CRACRAFT, 1980; WILEY, 1981). This should be considered an **apomorphic** character, as it is a specialisation that was absent in the immediate ancestor.

A problem often faced with the use of cladistic analysis stems from the experimental procedures required for the studies, which often make it impossible to use many species, allowing conclusions that do not reflect the real evolution of a given structure (KAAS, 2002). In practice, few species are available for observation. An alternative to the indiscriminate use of numerous species is based on comparing a restricted number of species chosen as representatives of a taxonomic group based on "quasi-phylogenetic" criteria (KAAS, 2002).

Based on this basic knowledge and concepts about the use of cladistic analysis in the study of evolution, we will define what characterises a cortical area, and then show which sensory-motor cortical areas would probably have been present in the ancestral mammal and which of them emerged or were lost in the species throughout the mammalian evolutionary process.

COMMON PLAN OF MAMMALIAN CORTICAL ORGANISATION

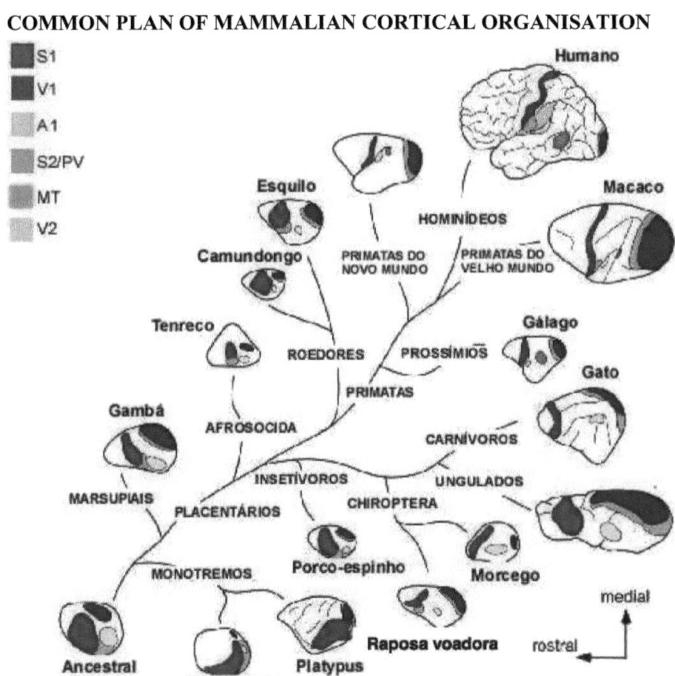

Figura 2. Phylogenetic tree illustrating the relationship between the cortical fields common to the three major groups of mammals. Modified from Krubitzer, 2007.

1.2 **Topographically organised cortical areas**

Sensory cortical areas can have a systematic representation of the sensory surface, as well as a motor representation of body movements. This representation is called a cortical map (KILLACKEY et al., 1995). With regard to these maps, we observe, for example, that there is a strong correlation between the sensory receptor periphery and its projection to the corresponding cortical region, which we call topographical organisation (KILLACKEY et al., 1995). In topographical organisations, the projections from a given group of peripheral receptors are grouped in an orderly fashion to the same target region on the cortical map, and can be observed in the projections from the thalamus to the cortex, in the projections between cortical areas, or between the cortex and subcortical structures (KILLACKEY et al., 1995).

Such topography is observed in retinotopic or visuotopic cortical maps (of the visual system), tonotopic maps (of the auditory system) and somatotopic maps (of the somatosensory system) (KILLACKEY et al., 1995). In these maps, there can be segregated processing of certain characteristics of the sensory stimulus, such as spatial localisation and sensory submodalities (KAAS, 1997).

Initially, several authors believed that topographical organisation was of little significance for cortical processing and incompatible with the associative functions performed by the neocortex (KAAS, 1997). Currently, the principle that topography is necessary for sensory discrimination prevails, since the spatial separation of neuron activity within topographic maps is important for localising and distinguishing separate stimuli on sensory surfaces (KAAS, 1997).

Neurons located in the same topographical region of the cortex usually communicate intensely during the processing of sensory or motor information (ALLMAN, 1999). Communication between functionally related neurons in a non-topographically organised system could result in longer and denser connections than in topographical organisations. In this way, it is suggested that the organisation of cortical areas in topographical maps could also represent an economical way of connecting neurons with similar functions (ALLMAN, 1999).

The first demonstrations of topographical maps in the cortex are ancient (ALLMAN, 1999). In 1740, Emanuel Swedenborg observed that certain fibres arising from the sensory organs connected indirectly with the cortex, and that other fibres arose from the cortex and went to the muscles of the body (ALLMAN, 1999).

Although it was ignored, Swedenborg already suggested in his time the existence of a topographical organisation for the motor cortex, with control of the foot muscles located dorsally in the cortex and the facial muscles in the ventral cortex (ALLMAN, 1999).

However, the subsequent ideas about cortical subdivisions into modules, proposed in 1810 by Franz Josef Gall and Johann Spurzheim, became better known (ALLMAN, 1999). They proposed that the encephalon was made up of organs responsible for personality traits such as humour, concentration, benevolence and tenacity. This theory was called phrenology (study of the mind) by Spurzheim, and although it was widely accepted at the time, it was not supported by any experimental data or clinical observation (ALLMAN, 1999; STREEDTER, 2005).

The direct identification of a topographical map in the cortex was only realised in 1860 by John Hughlings in epileptic patients (ALLMAN, 1999). He concluded, based on his clinical observations, that the muscles of the body were represented in the encephalon in a specific region. This map was confirmed in 1870 by the Germans Gustav Fritsh and Eduard Hitzig. They discovered the motor cortex by stimulating the surface of a dog's brain and observing the body movements generated by this stimulation. Other researchers also contributed to the description of motor maps in other animals, such as David Ferrier (responsible for the first publication of a cortical map) (ALLMAN, 1999).

The famous works by Charles Sherrington, Oscar Vogt and Wilder Penfield were the first to describe the magnification of the cortical representations of the hand and face in great apes and humans, showing that these parts of the body occupy a large area in the motor cortex (SCHIEBER, 2001; ALLMAN, 1999).

Later, in the 1930s, electronic amplifiers and oscilloscopes were introduced, allowing the electrical activity of the cortex to be recorded. The first electrophysiological recordings were made by Edgar Adrian and Clinton Woosley, in the cortical region adjacent to the motor cortex (ALLMAN, 1999). They named this region the somatosensory cortex (soma in Greek = body) because it was activated by mechanical stimulation of the body's surface. The cortical representation of a particular region of the body surface was identified by the successive movement of the electrodes across the cortical surface. In the 1970s, Jon Kaas, Michael Merzenich and their collaborators identified at least 4 cortical maps representing the body surface in the somatosensory cortex of primates (ALLMAN, 1999). Other cortical maps, this time related to the visual system, were also described in the 20th century. The first topographical map of the primary visual cortex was described by ophthalmologist Tatsuji Inouye during the First World War (ALLMAN, 1999). The topographical organisation of this map was suggested on the basis of the visual deficits presented by soldiers hit during the fighting. The region of greatest cortical representation observed by Tatsuji was the central region of the retina, to the detriment of the retinal periphery (ALLMAN, 1999). It wasn't until the 1960s that Jon Kaas and John Allman mapped the visual cortex using microelectrodes (with greater spatial resolution than surface electrodes) and observed that the visual system occupies a more extensive region in the cortex than initially observed and that it was made up of many other cortical areas (ALLMAN, 1999).

1.2.1 **Criteria for delimiting cortical areas**

As we saw in the previous section, the identification of cortical maps can be carried out using different methods, including electrophysiological recording and electrostimulation, as well as clinical observations. It is important to note that the criteria used to delimit a cortical area and a cortical map do not always lead to the same result. Each subdivision method has its weaknesses and, on its own, may not be conclusive in demarcating a particular area. Therefore, in order to delimit a cortical area, several study methods must be considered (KAAS, 1983).

Initially, analysing cortical architecture was the main method for delimiting cortical areas. One of the best known studies in this field was carried out by Brodman in 1909 on the cerebral cortex of mammals of all orders *(cf.* BRODMANN, 1909; STRIEDTER, 2005). However, other researchers have also described the architecture of cortical areas in different species and many disagreements have arisen regarding the borders they proposed (KAAS, 1987). Comparing architectural borders between species is a difficult task, since there is a great deal of variation in cortical morphology between different species (c/. BRODMANN, 1909). Furthermore, an architectural border that is obvious to one researcher may be considered by another researcher to be a variation within the area or a distortion caused by a groove. Another obvious problem is that the neocortex can be subdivided into several functional fields, and that architectural analysis only highlights some of these edges (KAAS, 1983).

To solve this problem, electrophysiological recording and stimulation techniques allowed for a more accurate delineation of cortical subdivisions. However, the first electrophysiological experiments only used surface electrodes, which revealed few details about the cortical maps (KAAS, 1983). Refined cortical subdivisions could only really be understood with cortical recording and stimulation using electrodes that were thinner (microelectrodes) than the previous ones, and capable of penetrating the entire cortical thickness.

In addition, each cortical area has a particular pattern of projections that can be observed using neuroanatomical techniques. These techniques can be used after the electrophysiological delimitation of an area or as an auxiliary technique in the characterisation of a cortical area through knowledge of its connections, as will be the case in our study (KAAS, 1983). The identification of the primary visual cortex (VI) is unison among the various methods of delimiting cortical areas. For this reason, the same criteria used to delimit VI serve as a reference for the other cortical areas (KASS, 1983). These criteria include:

1) a distinct cytoarchitecture (with a unique pattern of cellular organisation);

2) a representation of the contralateral half of the periphery (or sensory surface);
3) a unique pattern of connection with cortical and subcortical structures;

4) a single type of processing (sensory) and

5) specific deficits after injury or removal of the area in question.

1.3 Common plan of cortical organisation in mammals

Before understanding how and what changes have occurred in the encephalon over the course of mammalian evolution, it is necessary to bear in mind the similarities between the species studied so far. The cortical areas common to the three major branches of mammals (prototherians, metatherians and eutherians) were probably present in the common ancestor and form part of a common plan of cortical organisation in mammals (KAAS, 1983, 2004, 2008; KRUBITZER, 1995,2007).

The existence of a basic cortical plan is an important feature in the organisation of the neocortex. Although it is possible to modify the size, number, shape and internal organisation of some cortical areas, comparative neurobiology studies suggest that the complete elimination of some areas or the formation of a new plan would not be possible without altering other events involved in the formation of other cortical areas (KRUBITZER, 1995). An example of this is the presence of the visual cortex and the retino-geniculo-cortical pathway even in subterranean mammals with almost non-existent vision, such as the mole rat (KRUBITZER, 1995, 2007). These rodents have small eyes covered by skin, and their visual system is only used for circadian functions (KRUBITZER, 2007).

Along the length of the neocortex, specific sensory processing fields and one or more motor fields appear to be present in most mammalian groups (KRUBITZER, 1995, 2007; KAAS and COLINS, 2001b; KRUBITZER and KAAS, 2005). These fields include the primary visual, primary somatosensory and primary auditory cortex - VI, SI and Al, respectively (KRUBITZER, 1995, 2007; KAAS, 1983; KAAS and COLLINS, 2001b; KRUBITZER and KAAS, 2005). The proposed homology of these areas between existing mammalian species is supported by the similarities observed in the topographical organisation, architectural structure, connection pattern between the thalamus and cortex and preference for a certain type of stimulus (KRUBITZER, 1995).

Until recently, it was believed that these areas were the only ones preserved among mammals. However, the common plan of cortical organisation seems to be more complex than originally thought. Other areas have been described more recently in prototherians and metatherians, such as: the secondary somatosensory area (S2), ventral parietal area (PV), a rostral somatosensory field (R, DS or 3a), a motor field (M or Ml), an auditory field beyond Al and a secondary visual field (V2). Again, when compared in different species, these fields have the same topographical characteristics and the same pattern of connections and architectural organisation; this allows us to admit that they are homologous between the species studied and that they did not arise independently in the different lineages (KRUBITZER, 1995; KAAS and COLLINS, 2001b).

Figure 2 shows a cladogram of the main taxonomic groups of mammals and their representative species. It can be seen that primary areas VI, SI and Al are present both in mammals with smaller brains, such as the opossum, porcupine, mouse and prototherians (omitorrinco and echidna), and in mammals with complex brains, such as primates. It is possible to observe that, as the encephalon increases in size in some species, new cortical areas are "added" throughout the evolutionary process. This is what happens in primates. The middle temporal area (MT) is activated by moving visual stimuli (KAAS and COLLINS, 2001b), and can be seen in all the primates represented in figure 2, characterising it as a common feature of this group of mammals.

1.3.1 Somatosensory cortex

In several species of mammals, the somesthetic cortex is activated via the spinal thalamic pathway by stimulating superficial and deep cutaneous mechanoreceptors, as well as deep kinaesthetic receptors (proprioceptors located in tendons, joints and muscles) and visceroceptors (KAAS, 2004). This activation occurs through afferents coming from the sensory periphery, which enter the spinal cord and brainstem to synapse with a second-order neurone, which may be in the gracilis and cuneiform nuclei, or in the nuclei of the trigeminal pathway (KAAS, 2004). Spinothalamic information from lamina 5 neurons, nocireceptors, thermoreceptors and other neurons associated with interoception in lamina I, project to the VPi thalamic nucleus (CRAIG and ZHANG, 2006).

As illustrated in Figure 3, most mammals seem to have at least five somatosensory areas: Sl, S2, PV and two bands of somatosensory cortex located rostral and caudal to Sl (SR and SC, respectively) (KRUBITZER, 1995; KAAS, 2004; KAAS and COLLINS, 2001b). Even small-brained species, such as the opossum *Didelphis virginiana,* seem to have five somatosensory fields: Sl, SR, SC, S2 and PV (Figure 3) (BECK et al., 1996).

However, in some species, the PV area could not be distinguished from S2 (CATANIA et al., 1999). Areas SR and SC are often very narrow or even absent in some animals (CATANIA et al., 1999). These two areas are known to make dense connections with Sl, although they do not respond clearly to somatosensory stimuli in anaesthetised mammals (CATANIA, 2000). Only in primates are these responses more evident, allowing the characterisation of the areas rostral and caudal to Sl as area 3a and 1, respectively (Figure 3). In Old World primates, the somesthetic cortex has a well-characterised sixth area, caudal to area 1, called area 2 (Figure 3)

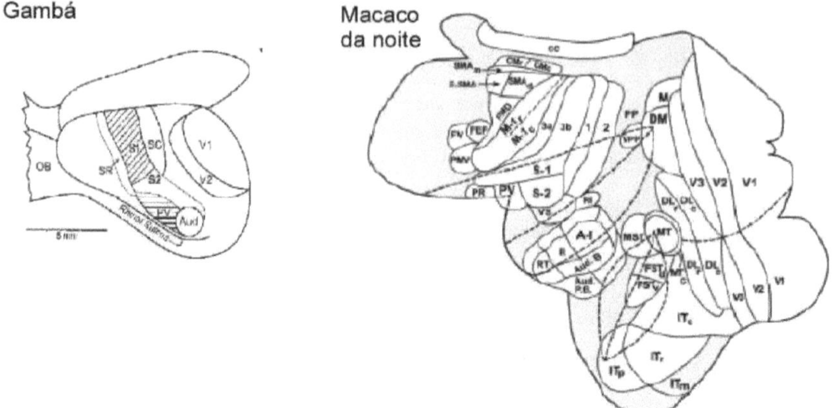

Figura 3.   Organisation of the somatosensory cortex in mammals: comparison between the opossum (marsupial) and the night monkey (primate). Note that the opossum's neocortex has a smaller number of cortical areas than the night monkey's neocortex (dorsolateral and flattened views, respectively). Modified from Kaas, 2004.

(PADBERG et al., 2005). The homology between the non-primate SC area and areas 1 and 2 is not yet well characterised.

Initially, the SI area was defined by electrophysiology as a single somatotopic representation of the body surface (*cf.* WOOLSEY and FAIRMAN, 1946; WOOLSEY, 1958). However, four architecturally distinct fields were later recognised in this region of the monkey: Brodman's areas 3a, 3b, 1 and 2 (Figure 3) (KAAS, 1983). According to Jon Kaas (1983), only area 3b in monkeys can be compared with area SI in non-primate mammals (KAAS, 1983). Both areas SI and 3b are characterised by 1) representing the cutaneous receptors of the contralateral body surface; 2) being the main target of projections from the posterior ventral nucleus (VP or VB) of the thalamus (the main somatosensory nucleus); 3) having a well-developed layer IV (KAAS, 2004).

1.3.2 Visual cortex

Although the number of visual areas varies greatly between mammals, all these species have at least area VI, and probably also area V2 (Figure 2). According to Rosa and Krubitzer (1999), rodents have an area V2 lateral to VI, rather than several smaller areas as was initially believed (*cf.* ROSA and KRUBITZER, 1999).

Furthermore, area V2 was recently described in a species of marsupial *(Dasyurus hallucatus),* which indicates its presence in a large number of mammals and possibly also in the common ancestor of mammals (ROSA, et al., 1999).

1.3.3 Auditory Cortex

The information obtained about the auditory cortex is not sufficient to characterise which areas are common to most species. As shown in Figure 2, only area Al has been consistently identified (KRUBITZER, 1995; KAAS and COLLINS, 2001b). However, the tonotopic representation of the primary auditory cortex does not seem to follow a common plan of organisation. In the guinea pig, the representation of low sound frequencies is located rostrally and those of high frequencies caudally (c/. WALLACE et al., 2000), while in gerbils this representation is organised in the opposite way *(cf.* BUDINGER et al. 2001b).

Some auditory areas are also known to be multisensory. In rodents, for example, the dorso-rostral secondary auditory field also responds to somatosensory stimuli *(cf.* WALLACE et al., 2000). Another secondary field, the caudomedial area (CM), also responds to both auditory and somatosensory stimuli *(cf.* SCHROEDER et al., 2001). No homology has been suggested for these areas, since there are not enough studies to conclude what is common to several mammals and what is a specific feature of a given species (KAAS and COLLINS, 2001b).

1.3.4 Multimodal areas

As can be seen in Figure 2, so far all the areas common to the three major mammalian groups are unimodal processing areas, i.e. they are involved in a single sensory modality.

In most mammals, a rudimentary form of posterior parietal cortex seems to be present, suggesting that this cortex may have arisen early in the evolution of the encephalon and remained in several species (PADBERG et al., 2005). The posterior parietal cortex is a large region of the cortex located between the somatosensory cortical fields rostrally and the visual fields located caudally (PADBERG and KRUBITZER, 2006). This region differs from the pure sensory areas and is considered a high-hierarchy specialisation of the sensory-motor system for the following reasons: 1) they show responses to more than one sensory modality, generally having neurons that can respond to both somesthetic and visual stimuli; 2) microstimulation of some areas of the posterior parietal cortex of primates induces defensive, aggressive and object-reaching movements (STEPNIEWSKA et al., 2005; COOKE et al., 2003); 3) these areas are intensely interconnected with somesthetic and visual sensory areas, and with frontal motor fields (PADBERG et al., 2005; *cf.* RIZZOLATI and FADIGA, 1998).

According to Kaas (1987), the addition of new cortical areas throughout the evolution of the brain seems to have been based on the addition of new unimodal areas, and not by the addition of the multimodal cortex. More studies are needed to confirm the presence of a multimodal area homologous to the posterior parietal cortex in all mammalian groups and to trace a line of evolution for this region.

1.3.5 Motor areas

The presence of a primary motor cortex (Ml) is consistent among placental mammals (KAAS, 2004; KARLEN and KRUBITZER, 2007). In these mammals, area Ml is characterised by: 1) thalamic projections originating mainly from the ventral lateral nucleus (VL); 2) cortical projections from S2/PV, SR, SC; 3) Betz or giant pyramidal cells in layer V and the absence of an obvious TV layer, characterising it as agranular cortex; and 4) a map of movements from the face to the hind limb evoked by electrostimulation (KAAS, 2004).

However, in non-placental mammals the presence of a primary motor cortex with these characteristics is questionable. In marsupials, for example, area Ml has not been localised and some authors suggest that the somatosensory cortex is responsible for motor functions (see next items) (KAAS, 2004; BECK et al., 1996; KARLEN and KRUBITZER, 2007). In some prototherians, such as echidna, area Ml seems to occupy a separate region from SI *(cf.* LENDE, 1964; ULINSKY, 1984; KRUBITZER, 2005). However, in omitorrinco, another prototherian, the first electrophysiological studies suggested a partial overlap of this area with SI (BOHRINGER and ROWE, 1977). The presence or absence of Ml in the ancestral mammal will be discussed below.

1.4 Changes that can occur in the brain of different mammal species

Although we can propose with relative certainty a common plan of cortical organisation among mammals, a detailed analysis of the neocortex of various species allows us to establish some differences in this cortical organisation. These variations are generally associated with the various morphological and behavioural specialisations presented by each species in its evolutionary process (KRUBITZER, 1995, 2007; KRUBITZER and KAAS, 2005; KARLEN and KRUBITZER, 2007; CATANIA, 2005).

Figure 4 illustrates the main phenotypic variations observed between different mammal species (KRUBITZER, 2007):

1) changes in the size of the neocortex;

2) changes in the size and shape of cortical areas;

3) changes in the internal organisation of a cortical area;

4) appearance of modules in a cortical area;

5) appearance of new cortical areas;

6) changes in the connection patterns of cortical areas.

In the following sections we will describe some theories involving the modifications and evolutionary mechanisms of cortical areas.

Possible changes to the neocortex

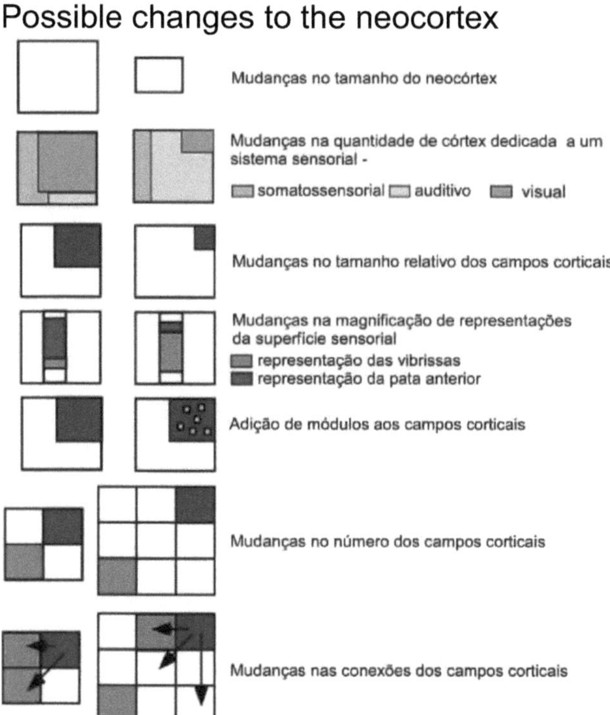

Figura 4. Types of changes that have occurred in the mammalian neocortex. Modified from Krubitzer, 2007.

1.4.1 Changes in the size of cortical areas

Cortical areas have undergone changes in size during the evolutionary process of each species. Two main hypotheses have been discussed regarding how these brain areas individually changed their size over the course of vertebrate evolution (STRIEDTER, 2005).

Considering that different species of animals exhibit specific behaviours and that certain brain areas are related to these different behaviours, it is possible that throughout evolution, brain regions have specialised independently in order to reflect the specialised behaviour of the species. This hypothesis is called ***mosaic evolution*** (STRIEDTER, 2005).

However, it cannot be ignored that brain areas are interconnected, forming a complex processing network, which indicates that changes in the size of one area can cause changes in other areas, what we call evolution in *concert* (KRUBITZER, 2007; STRIEDTER, 2005).

Some physiological studies have shown correlations between increased size of brain areas and highly developed sensory abilities. The mechanism by which larger areas result in better processing of sensory and motor information was first defined by Harry Jerison in 1973 in the *principle of mass property* (STRIEDTER, 2005).

This mass property principle suggests, from comparisons between species, that the importance of a function in the life of each species can be reflected by the absolute amount of neural tissue for the function in the species. It also implies that, within a species, the relative mass of neural tissue associated with different functions is related to the importance of this function in the species *(cf.* JERISON, 1973, p. 8-9).

If we consider the absolute size of certain regions, this statement may not be correct. Smaller regions of the encephalon may have less information processing, but they can still play an important role in an individual's life (STRIEDTER, 2005). The suprachiasmatic nucleus, for example, is one of the smallest cell groups in relation to the absolute size of the encephalon. However, because it is fundamental in maintaining the internal "biological clock", the size of this nucleus is highly conserved between species and seems to be important in the life of the organism (STRIEDTER, 2005).

An alternative to analysing the absolute size of brain structures is to analyse their proportional size, i.e. how large a structure is compared to the rest of the encephalon or neocortex (KRUBITZER, 1995, 2007; KRUBITZER and KAAS, 2005; CATANIA, 2005). This comparison, when made between different species, can indicate the importance of a region in the behaviour of the species.

Based on the concept of relative size, we can mention the cortical changes that occur in the size occupied by a sensory system in the cortex of a cortical area and in the internal organisation of this area (Figure 5). The three species mentioned in the figure (mouse, opossum and bat) have approximately the same size cortical surface, but the amount of cortex devoted to a particular sensory system varies according to the use of certain sensory receptors by the species (KRUBITZER and KAHN, 2003).

The omitorrhine is another classic example where specialisations of peripheral morphology seem to influence the magnification of cortical areas and the specialisation of cortical representations (Figure 6). The SI area of this animal not only occupies a proportionally larger region in the neocortex than areas VI and Al, but is also proportionally larger than in other mammals (KRUBITZER, 1995, 2007). This expansion of SI in the omitorrhine is related to the density of electroreceptors and mechanoreceptors located in the beak. Cortical magnification of other somatosensory areas of the omitorrhine, such as areas R (rostral somesthetic) and S2/PV, reinforces the importance of the somatosensory system in this species. Nevertheless, the cortical representation of the beak occupies almost 90 per cent of area S1, as shown in Figure 6.

1.4.2 Changes in the number of cortical areas

At the beginning of evolutionary neuroscience, it was believed that the brain evolved and became more complex by adding new structures. This belief began with Ludwig Edinger in 1908 and was widely accepted in the first half of the 20th century (STRIEDTER, 2005). Edinger proposed that the evolution of the vertebrate telencephalon occurred in progressive stages with increases in complexity and size, culminating in the human encephalon (THE AVIAN BRAIN NOMENCLATURE CONSORTIUM, 2005). According to his proposal, the pallencephalon was the old brain, involved in instinctive behaviour. The addition of new structures would have formed the neotelencephalon (pallium or cortex), which would control learning and intelligence. These sub-divisions of the telencephalon became widely known by the prefixes paleo (oldest), archaic (archaic) and neo (new) and designated the evolutionary order of a given structure (THE AVIAN BRAIN NOMENCLATURE CONSORTIUM, 2005).

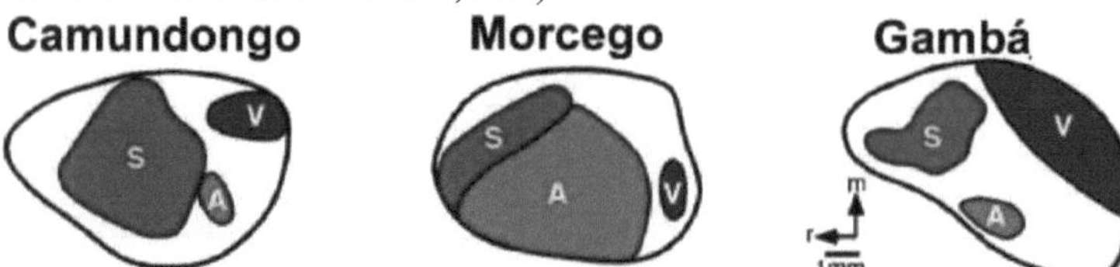

Figura 5. Proportional size of sensory systems in three species with similar neocortex size. Representation of a tangential view of the cortex of the mouse, bat and opossum. Note that the proportion of area occupied by the sensory systems in the neocortex varies between species. Modified from Krubitzer and Kahn, 2003.

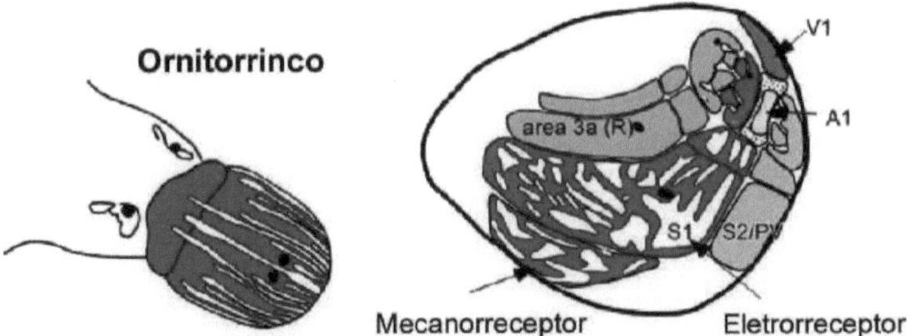

Figura 6. Cortical magnification of body parts important to the animal's behaviour. The cortical representation of the omitorrhinx's beak occupies most of SI (in red). Areas VI and AI occupy a proportionally smaller region in the neocortex of this species. Modified from Krubitzer, 2007.

The pallium of fish was then called the palaeocortex and was the ancestor of the human olfactory cortex. Reptiles would have added an archicortex, which would have been the predecessor of the human hippocampus. While in birds no addition of structures was initially proposed, in mammals the appearance of the neocortex has been suggested, responsible for their more sophisticated behaviour (THE AVIAN BRAIN NOMENCLATURE CONSORTIUM, 2005).

On the other hand, J.B. Johnston stated, based on his research and the literature, that regions considered new or unique to a mammal species by Edinger had homologues in fish and reptiles (JOHNSTON, 1923). Based on this evidence, Johnston stated that new structures do not appear in the encephalon "out of nowhere" during evolution. Following this reasoning, the term isocortex is currently suggested as a reference to the cortex characteristic of mammals (NORTHCUTT and KAAS, 1995, STRIEDTER, 2005), rather than neocortex (a completely new cortex). However, the term neocortex continues to be used in the literature and seems to be correct when referring to the exclusivity of this structure in mammals, although it does not mean that it is a cortical organisation that evolved completely independently of the paleo and/or archicortex (THE AVIAN BRAIN NOMENCLATURE CONSORTIUM, 2005).

If we compare the neocortex of different mammals, we can clearly see that different species differ in the number of cortical areas devoted to a cortical system. So we ask again: how did the encephalon become more complex over the course of evolution?

The most widely used explanation for this question is that certain regions of the ancestral encephalon were subdivided into two or more regions in the descendant encephalons (STRIEDTER, 2005). Thus, the regions that appeared in the encephalon would be called "new" because they were not present in the ancestor, but were not "added" to the old structures from scratch.

These regions may have appeared through two different mechanisms: *phylogenetic conversion* and *proliferation* (STRIEDTER, 2005). In *phylogenetic conversion,* a region

The brain is transformed into a new region through structural changes, such as those that occur in connectivity, cytoarchitecture or histochemistry (STRIEDTER, 2005). In this mechanism, there is one-to-one homology between the character in all species (Figure 7A). However, one character (the "new" one) is different from its homologues.

A classic example of this *phylogenetic conversion is* the emergence of the neocortex in mammals from a poorly laminated cortical structure in other vertebrates. In this case, there is a one-to-one relationship between the precursor cortical structure and the neocortex.

On the other hand, *phylogenetic proliferation* occurs when one species has more characters than another and these characters are not homologous to the ancestor (STRIEDTER, 2005). There are two distinct mechanisms accepted for the occurrence of phylogenetic proliferation: *segregation* and *addition* (Figure 7B). The term *segregation is* often used to describe the evolution of homogeneous regions in the ancestor into multiple, distinct regions that are not homologous to the ancestor (STRIEDTER, 2005).

Evolution of new cortical areas

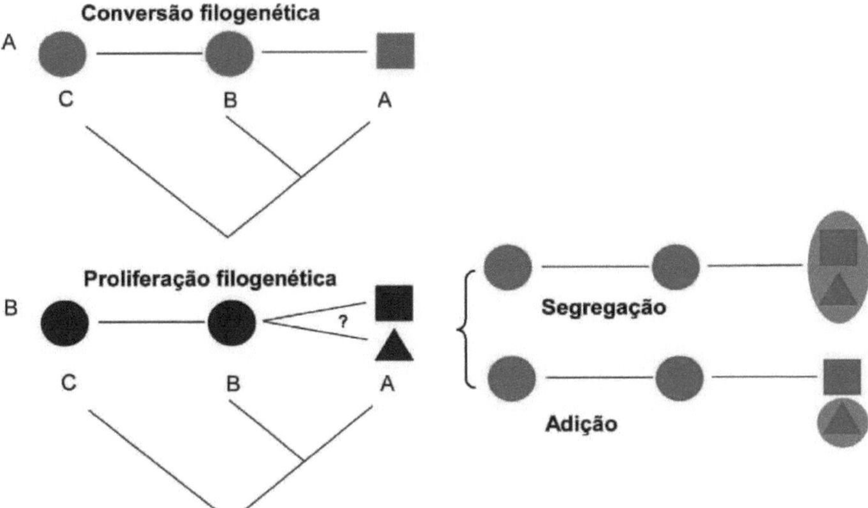

Figura 7. Schematic drawing of phylogenetic conversion and proliferation. (A) illustrates phylogenetic conversion where a single character is different from its homologue. New cortical areas can also emerge and not necessarily show a relationship of homology with the analogous ancestral structure, as in the case of phylogenetic proliferation (B). Modified from Striedter, 2005.

The hypothesis of evolution by segregation is exemplified by Ann Butler (1994), who explains the greater number of nuclei in the dorsal thalamus of amnionts than in anamnionts. According to Butler, the dorsal thalamus is subdivided into two groups of nuclei: the lemnothalamus and the cholothalamus (STRIEDTER, 2005). The lemnothalamus in anamnionts, fish and amphibians consists of a single nucleus, the anterior nucleus. Meanwhile, in amnionts, including mammals, several nuclei are observed in this region. It is known that none of these nuclei in amnionts is homologous in isolation to the anterior nucleus of anamnionts, and it is therefore assumed that this is an example of *phylogenetic segregation* (STRIEDTER, 2005). Thus, if these nuclei are not homologous, it cannot be said that there has been a conversion of one ancestral nucleus into several distinct nuclei, as occurs in *phylogenetic conversion.*

In general, when the segregation and conversion hypotheses do not satisfy the explanations for the appearance of structures in the encephalon, *phylogenetic addition is used* as an alternative hypothesis. In this model, new structures would be added with the appearance of new genetic expressions from an embryologically precursor region (STRIEDTER, 2005). The modified precursor could then generate a new structure in the adult. Another type of *phylogenetic addition* could occur when the embryologically precursor region expands. This expansion would leave a part of the cortex unexposed to the old developmental stimulus, thus generating a new developmental trajectory (STRIEDTER, 2005).

But what about the cortical areas?

An argument in favour of the hypothesis of phylogenetic segregation as a mechanism responsible for the appearance of **new areas** is strongly supported by Richard Lende's (1969) proposal that the primary somatosensory and motor areas correspond to a single cortical area in ancestral mammals, forming a sensory-motor amalgam.

According to Lende (1969),

"(...) the present pattern of localisation of cortical areas evolved from a polymodal pallium containing completely overlapping representational areas (e.g. sensory and motor)." (LENDE, 1969, p. 273).

As for other species, Jon Kaas and collaborators mapped the sensory cortex of various mammals and found at least the primary and secondary sensory areas to be well segregated in these species (KAAS, 1987; KRUBITZER, 1995; NORTHCUTT and KAAS, 1995; CATANIA et al., 1999). Fewer studies have been carried out on the motor cortex, but most species seem to have a primary motor cortex and some premotor regions. Thus, it is suggested that these areas may have separated during the evolutionary process of mammals from an amalgamation in the common ancestor. It remains to be seen, however, whether these areas really arose from a polymodal cortex in the ancestor as Lende (1964) suggests, which we will discuss below.

1.4.3 Changes in neural connections

Changes in the number of cortical areas over the course of evolution consequently imply a rearrangement of the connections between these areas and the rest of the cortex (STRIEDTER, 2005). Or, changes in neural connections during the evolution of a species could lead to the appearance of new cortical areas (STRIEDTER, 2005).

Initially, C. J. Henick's *invasion theory* predominated among neuroanatomists. According to this theory, unusual connections could invade a region and favour cortical differentiation (STRIEDTER, 2005).

In line with this hypothesis, Leah Krubitzer (1995) proposes that new cortical areas arise from the invasion of new afferent thalamic projections, causing a realignment of pre-existing projections (Figure 8). Thus, the new projections could form cortical modules, aggregating within the areas. Throughout the evolutionary process, some fields could generate a complete segregation between the "new and old" projections, or the new field with both projections could remain within the invaded area, generating new cortical areas (KRUBITZER and KAHN, 2003).

In 1980, Ebbesson proposed the *parcellation hypothesis in an* attempt to explain these changes in cortical connections (Figure 9A).

According to Ebbesson,

"The nervous system has become more complex, not by one system invading another, but by a process of parceling out that involves the selective loss of connections between neighbouring systems and aggregates formed" (c/. EBBESSON, 1980, p. 213).

Richard Lende's (1963) sensory-motor amalgam is a commonly cited example of Ebbe ssonian parcellation (Figure 9B). Studies with neurotracers show that both the somesthetic and motor thalamic nuclei project to the sensory-motor amalgam region in the marsupials studied (DONOGUE and EBNER, 1981). In placental mammals, the primary somatosensory and motor areas are spatially separated and the somesthetic and motor nuclei of the thalamus project respectively to these areas without overlapping projections (KAAS, 2004).

1.5 Sensorimotor amalgam hypothesis

The sensorimotor amalgam hypothesis was first described by Richard Lende in 1963. At the time, there were many doubts about the somatotopic representation of the marsupial motor cortex. Few studies prior to Lende's were able to identify the motor representation of the regions caudal to the anterior paw, such as the tail and posterior paw, in the neocortex (KARLEN and KRUBITZER, 2007).

In 1940, the primary motor cortex of six species of marsupials *(Perameles nasuta, Sacrophilus harrissi, Dasyurus viverrinus, brush-tailed opossum, Macropus rufogriseus, Macropus agilis)* was studied by Abbie.

In none of these was a motor representation of the animal's entire body found (KARLEN and KRUBITZER, 2007).

Based on his studies, Abbie (1940) stated that the data was inconclusive on the existence of a complete somatotopic representation of the body in the motor cortex of *Didelphis virginiana* (a didelphid) (KARLEN and KRUBITZER, 2007). In 1963, Richard Lende used the most refined methods of the time to map the motor cortex in detail in this species. Using electro-stimulation with bipolar electrodes, he found the representation of the hind paw and tail in the medial wall of the neocortex. In addition to the motor cortex, the primary somatosensory cortex was also mapped using the evoked potential technique (LENDE, 1963a,b).

However, Lende's main result was that the primary motor cortex of the opossum was completely overlapping with the primary somatosensory cortex in the parietal cortex, which he called a *sensory-motor amalgam* (Figure 10).

According to Lende (1963),

"The somatic motor representation in the opossum's cerebral cortex is spatially coincident with the sensory somatic representation. There is no homologous area in the position of the precentral motor area. This sensory-motor area is not divided into a more sensory portion and a more motor portion. The localisation pattern within this area of both sensory and motor aspects corresponds to that found in the primary somatosensory area of placental mammals."

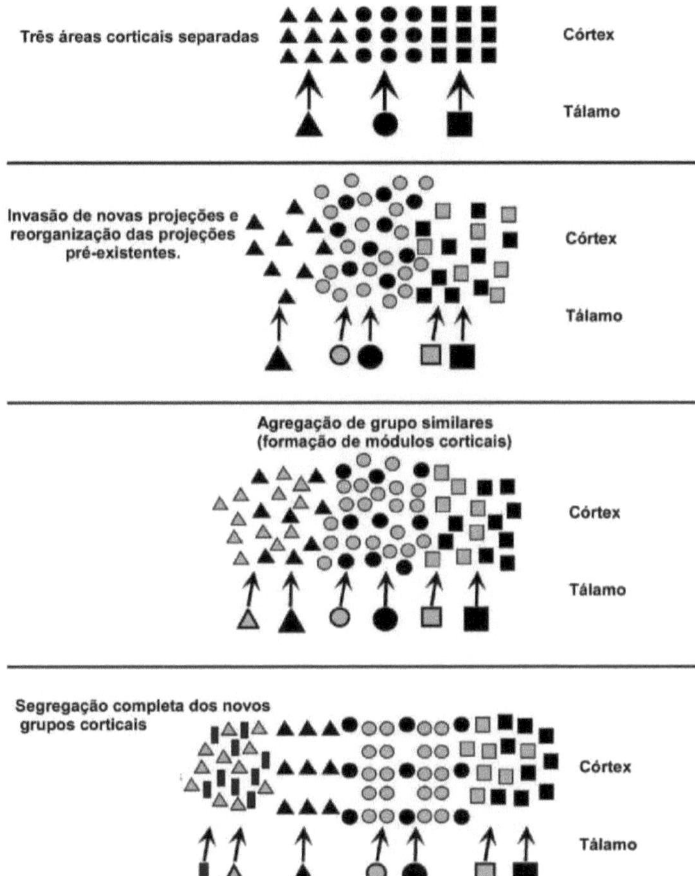

Figura 8.   Modification of neural connections. According to the invasion theory, new cortical areas could emerge throughout the evolutionary process due to the invasion of new neural connections. Adapted from Krubitzer and Kahn, 2003.

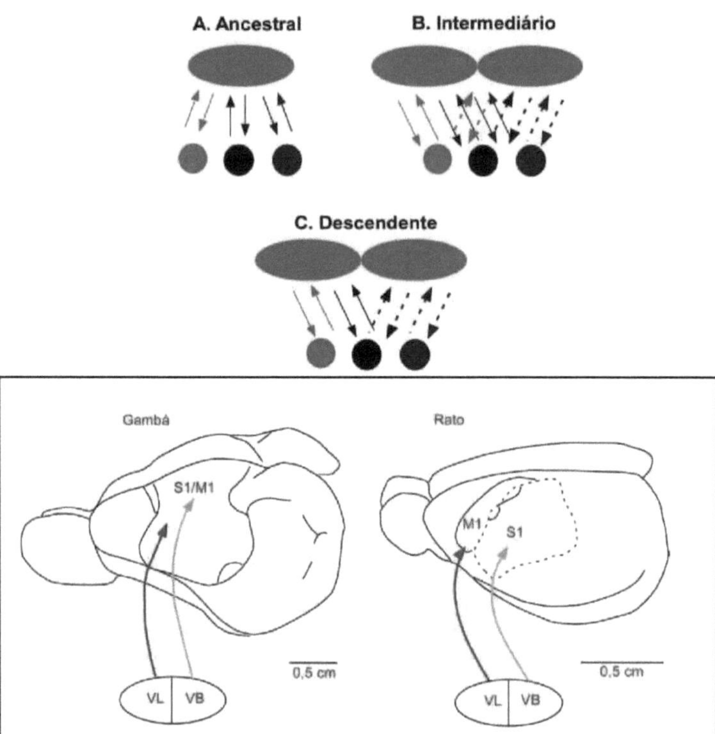

Figura 9.   Partitioning of cortical areas and connections. (A) shows a possible ancestral area (above) with its respective thalamic connections (below). In the Ebbessonian parcellation hypothesis (B), the cortical areas and their connections duplicate to form an intermediate condition. Over the course of evolution, these connections could segregate, generating new patterns of projections for both areas (C). Similarly, it is suggested that M1 would have evolved from a sensory-motor amalgam in the ancestral mammal (represented in the box inside the figure by the opossum brain) and that the convergence of the motor (VL) and somesthetic (VB) thalamic projections to Sl/M1 would have segregated in existing mammals (represented in the box by the drawing of the rat). Modified from Striedter, 2005.

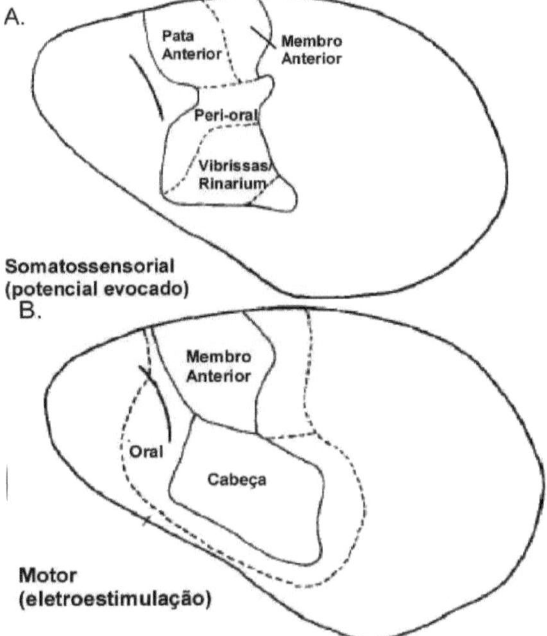

Figura 10. **Sensory-motor amalgam hypothesis.** Dorsolateral view of the neocortex of the opossum *Didelphis aurita* illustrating the superimposition of the somatosensory (A) and motor (B) maps in the parietal cortex of this animal. Modified from Pubols et al., 1976.

The proposal that the opossum's sensorimotor amalgam could be a primitive feature of the mammalian isocortex stimulated new studies in the 1970s aimed at confirming the organisation of the primary motor and somatosensory areas in these animals.

In 1976, Pubols and colleagues remapped the parietal cortex of *Didelphis virginiana* using microelectrodes (PUBOLS et al., 1976). These electrodes, unlike the surface electrodes used by Lende (1963), penetrated the cortex, recording a reduced volume of cortical tissue and allowing greater spatial resolution of the cortical somatotopic representation. The neuronal properties and receptor fields of SI were similar to those of other mammals, but the body surface appears to occupy a smaller portion of the parietal cortex than previously proposed by Lende (1963).

To verify the existence of overlapping sensory-motor systems in another marsupial species of the didelphid order, a new mapping of motor and somatosensory representations was carried out in the opossum

Didelphis azarae azarae (MAGALHÃES-CASTRO, 1971). Again, the occupation of coincident areas by the motor and somatic sensory systems in the parietal cortex was demonstrated.

Subsequently, a study of the somatosensory cortex of the opossum *Didelphis virginiana* presented topographical, architectural and hodological (connections) evidence for the existence of five somatosensory areas in this species (BECK et al., 1996). Three representations of the body surface were found in the parietal cortex. These representations correspond to areas Sl, S2 and the ventral parietal area (PV). Two somatosensory fields were additionally proposed, for which, however, no somatotopic representation was identified. The new fields found are located in the region rostral and caudal to Sl, which were named the rostral somatosensory area (SR) and the caudal somatosensory area (SC).

The latter were considered to be separate areas from Sl, because they did not show a clear response to somatic stimulation in the anaesthetised preparation and because of the pattern of topographical projections from Sl to these somesthetic areas (BECK et al., 1996).

In addition, the silver impregnation technique in the myelin sheath (GALLYAS, 1979), widely used to delimit cortical areas, was useful in identifying the Sl, S2 and PV areas, which were densely marked (BECK et al., 1996). The SR and SC areas were also visible by this method, but were not as myelinated as the other somatosensory areas, and were therefore less marked by this method.

Microstimulation of the Sl and surrounding areas was carried out in the same study (BECK et al., 1996). In monkeys and other mammals, currents of 5 to 20 pA applied to the motor cortex are capable of producing movements (STEPNIEWSKA et al., 1993). In the opossum, surprisingly, no movements were observed at most of the stimulation sites, even with the use of currents above 100 pA, thresholds much higher than those found in other species (BECK et al., 1996).

Another contribution to clarifying the ancestral condition of the parietal cortex in mammals consisted of motor and somatosensory mapping in the South American marsupial *Monodelphis domestica* (FROST et al., 2000). According to the authors, this species is closer to the first metatherians than the *Didelphis*. This study used multi-unit recording with microelectrodes to identify the Sl and S2 areas. Electrostimulation was carried out using three different methods: intracortical microstimulation, deep low-impedance stimulation and superficial bipolar stimulation. This last method was similar to the one carried out by Lende (1963). In all three

30

procedures, the movements obtained were restricted to the facial region, and were superimposed with the representation of the mandible, vibrissae and snout in Sl. However, these results were obtained with currents of up to 60 pA, again much higher than the current needed to evoke movement in the motor cortex of other mammals.

1.5.1 Thalamo-cortical projections for the region of the proposed sensorimotor amalgam

The thalamus is a nuclear mass located in the diencephalon, which has various cell groupings with different anatomical and functional organisations (MOUNTCASTLE, 1980). Some thalamic nuclei are associated with systems for processing signals from the outside world and others participate in somatic and autonomic motor functions. The interconnection of the thalamus with the cortex forms an important anatomical substrate for our perception of the external world and the body, as well as our actions and motivational responses (MOUNTCASTLE, 1980).

When we analyse the connections of the somatosensory areas, we see that the axons related to cutaneous receptors (superficial and deep) are directed mainly to the posterior ventral thalamic nucleus (VP or VB) (KAAS, 2004). Information from deep receptors, such as neuromuscular spindles, at least in primates, is directed to the ventral posterior superior nucleus (VPS). From these nuclei come the projections directed to the somesthetic areas located in the anterior parietal cortex.

Cortical motor areas are reciprocally connected with specific thalamic nuclear groups. Such thalamocortical projections are critical for the functioning of the motor cortices (KAAS and STEPNIEWSKA, 2000). The efferent cerebellar nuclei, the internal segment of the globus pallidus and the substantia nigra are subcortical motor centres that receive indirect information from a variety of somatosensory, motor and associative cortical areas and send this information back to the motor cortex via the thalamus (KAAS and STEPNIEWSKA, 2000). Throughout our study, we will refer to the thalamic nuclei that connect the subcortical motor centres with the motor cortex as the motor thalamic nuclei.

The opossum's motor nuclei are the ventral lateral (VL), ventral anterior (VA) and ventral medial (VM) thalamic nuclei. The VL, VA and VM nuclei receive cerebellar projections in *Didelphis virginiana* (WALSH and EBNER, 1973). In primates, the ventral lateral nucleus (VL) also receives projections from the deep nuclei of the cerebellum *(cf.* JONES, 1985). The cerebellar information that reaches VL goes to the primary motor

31

and pre-motor cortices.

In order to verify the nature of the information arriving in the opossum's sensory-motor amalgam region, thalamic projections to the parietal cortex were observed in the opossum *Didelphis virginiana* using the anterograde degeneration technique in the 1970s (KILLACKEY and EBNER, 1973). Lesions using direct current were made in the thalamus in three different nuclei: VP, VAL (VA and VL) and the central intralaminar nucleus (CIN). The degenerated axons and axon terminals showed that the projections of these three thalamic nuclei converge on the entire parietal cortex. In this species, the VP nucleus receives projections from the medial lemniscus while the VAL nucleus receives from the cerebellum. The CIN nucleus is located between the medial and lateral subdivisions of the thalamus and receives convergent projections from the medial lemniscus, the cerebellum, the spinal cord and the reticular formation (DONOGHUE and EBNER, 1981; WALSH and EBNER, 1973).

With the emergence of neuroanatomical techniques with greater spatial resolution, the projections of the thalamic nuclei to the parietal cortex were re-evaluated. In 1981, Donoghue and Ebner carried out several injections of horseradish peroxidase (HRP) in this region. This tracer is taken up by the axons at the injection site and transported retrogradely to the cell body, allowing the origin of the thalamocortical projections to be detected. These authors performed injections in the following cortical regions: 1) the post-orbital cortex (i.e. between the orbital fissure and the anterior border of SI), in 2) SI proper, and 3) in the posterior parietal cortex (identified as the region of the parietal cortex immediately caudal to SI). As a result, it was again observed that, in SI, the projections originated in the VB, VL, CIN nuclei and the lateral central nucleus (CL). The CL nucleus is located anterior to the CIN nucleus and receives projections from the spinal cord, cerebellum and pretectal nucleus (DONOGHUE and EBNER, 1981).

The markings resulting from the post-orbital injection were mainly in the intralaminar and medial nuclei. The injection of HRP into the posterior parietal cortex marked only the VL nucleus, between the ventral nuclei of the dorsal thalamus. Thus, it was concluded that the cortex corresponding to Lende's (1963) somatomotor amalgam receives convergent projections from the VB "somesthetic" nucleus and the VL "motor" nucleus, without any segregation between them (DONOGHUE and EBNER, 1981).

An important reservation in relation to this study is the lack of delimitation of somesthetic areas in the anterior parietal cortex, which until then had not been identified. It is possible that these rostral and central injections carried out in the Sl area by Donoghue and Ebner (1981), do not only correspond to Sl, but also include part of the SR and SC areas described later by Beck et al (1996) (which we will describe in the next paragraph). Furthermore, in this same study only one neurotracer was used, preventing a comparison of injections in the same specimen. HRP also has the disadvantage of greater scattering and lower resolution than the tracers currently used (conjugated to dextran, for example) (REINER et al., 2000; VERCELLI et al., 2000).

As we saw in the previous section, the results found by Lende (1963) on the localisation of the primary motor cortex overlapping with the primary somesthetic cortex were not confirmed later in the opossum by Beck et al (1996) nor by Frost et al, 2000. However, the "motor" thalamic nuclei seem to project to the region of the amalgam, reinforcing the idea of motor processing in this region. It is now known that this region has further subdivisions that include SI, SR and SC (BECK et al., 1996). Projections to the SR and SC have not been studied in the opossum so far.

1.6 The opossum *Didelphis aurita* **as an experimental animal**

The main objective of our work is to analyse the organisation of the primary somatosensory cortex in the opossum *Didelphis aurita* by characterising the thalamocortical and corticocortical projections of areas SI and SC, as described by Beck et al. (1996).

Skunks are marsupials representing the metatherians, one of the three subclasses of mammals: prototherians (monotremes), metatherians (marsupials) and eutherians (placentals). In general, the term marsupial is associated with mammals that have a pouch or marsupium, but it should be emphasised that not all species in this group have a marsupium (KARLEN and KRUBITZER, 2007). The term metatherian, on the other hand, is more appropriate, as it means that marsupials represent a middle ground between prototherians (whose development does not occur in the placenta) and eutherians (which have a placenta during gestation).

It can be said that marsupials form a very diverse group, with approximately 270 species divided into seven different orders (KARLEN and KRUBITZER, 2007). Consequently, their representatives occupy different *habitats* and exhibit different behaviours. We find terrestrial, arboreal or aquatic species; very sociable species *(Dactylopsila trivirgatã),* others solitary like the koala *(Phascolarctos cinereus).* Marsupials can also be carnivores, herbivores or omnivores (KARLEN and KRUBITZER, 2007).

As a result of this biological diversity, we see marsupials with different morphological adaptations, which are also reflected in differences in brain size, cortical organisation and brain connectivity. (KARLEN and KRUBITZER, 2007). In the wild, marsupial brains range from small, smooth and with a small proportion devoted to the neocortex *(Didelphis virginiana)* to large, gyrencephalic brains with a proportionally large neocortex *(Vombatus ursinus)*. (KARLEN and KRUBITZER, 2007).

These diversities in morphological and behavioural characteristics make the marsupial group important for studying the relationship between the brain and animal behaviour. Many of today's marsupials have maintained a cortical organisation similar to their ancestor and are more similar to the ancestor of mammals than other current mammals. For this reason, theories about the evolution of cortical areas are based on the results obtained from the study of marsupial brains (KARLEN and KRUBITZER, 2007).

In this study, the opossum *Didelphis aurita* was chosen according to "quasi-phylogenetic" criteria (KAAS, 2002) as a model for studying the basic organisation of the sensory-motor cortex in mammals. The opossum *Didelphis aurita* is part of the didelphid order, which represents the American opossums and is the oldest of the marsupials (KEMBLE, 1997). These marsupials have a small brain in relation to their body and a cortical organisation with few areas (KARLEN and KRUBITZER, 2007). The small brain is similar to that of reptiles and the first mammals, which were also smooth and remained so until 50 to 60 million years ago (BECK et al., 1996).

However, based on the cladistic method, information on the organisation of the somatosensory cortex of species from other mammalian groups (Eutherians and Prototherians) was also considered in our study in order to infer the cortical organisation of the sensorimotor system in the common ancestor of mammals.

CHAPTER 2

OBJECTIVES

2.1 General Objectives

2.1.1 To see if the sensory-motor system of the opossum *Didelphis aurita,* a species with a primitive brain, has somatosensory areas similar to those of mammals with a complex neocortex or if it represents a simplified organisation of this system.

2.1.2 We then evaluated how the somatosensory area of the opossum *Didelphis aurita* could represent a stage in the parceling of connections proposed by Ebbesson (1980), observing whether or not somesthetic and motor thalamic projections converge on the areas studied.

2.2 Specific objectives

2.2.1 To study the thalamo-cortical projections to the SI and SC areas, especially the VL, VA and VM motor nuclei, and the VB somesthetic nucleus. The thalamo-cortical projections to the area immediately anterior to SI have been explored previously in this student's master's thesis (ANOMAL, 2004) and by Santiago Martinich (MARTINICH et al., 2000).

2.2.2 To this end, we will first confirm the architectural segregation of the motor and somatosensory thalamic nuclei using cytochrome oxidase, NADPH-diaphorase, calbindin and parvalbumin techniques.

2.2.3 To study the cortical projections for the SI and SC areas.

2.2.4 To achieve this, we will identify the precise borders of the SI and SC areas using electrophysiological and architectural methodologies (myeloarchitecture) at the same time.

CHAPTER 3

MATERIALS AND METHODS

3.1 Animals

The animal chosen as the experimental model was the *Didelphis aurita* opossum. This animal is a wild marsupial found in the forests of Rio de Janeiro. The animals used were donated by the IBAMA Screening Centre (licenses no. 20/06-CT-RJ, 61/06- CT-RJ) or captured in the Atlantic forests of Guapimirim - Rio de Janeiro (IBAMA license no. 87/05-RJ). These animals remained in the vivarium of the Carlos Chagas Filho Biophysics Institute at the Federal University of Rio de Janeiro until the time of surgery. All the experimental procedures and protocols described below were approved by the Committee for the Evaluation of the Use of Animals in Research at the Carlos Chagas Filho Biophysics Institute (CAUAP protocol no. 053-06-2005).

Of the 19 opossums used in total, we were successful with 8 adult opossums, of both sexes and weighing between 1.4 and 2.8 kg. Of these animals, 2 were used exclusively for the architectural analysis of the thalamus and cortex, while the other six were used either for electrophysiological experiments or for the injection of neurotracers (Table 1).

	Mapping Electrophysiological	Injection Neurotracer	Thalamic Architecture
Case 0409	-	Yes	-
Case 0506	Yes		Yes
Case 0602	Yes	Yes	Yes
Case 0629	Yes		Yes
Case 0630	-	-	Yes
Case 0631	-	■	Yes
Case0701	Yes	Yes	Yes
Case 0705	Yes	Yes	Yes

Table 1. Summary of the procedures carried out in each case. In all the neurotracer injection experiments analysed, the thalamic nuclei were identified using Nissl and/or cytochrome oxidase staining. In cases 0630 and 0631, the thalamic nuclei were identified using Nissl, cytochrome oxidase, NADPH-diaphorase, calbindin and parvalbumin.

3.2 . General procedures

The electrophysiological mapping and neurotracer injection experiments involved two stages, separated by a survival interval of 10 to 12 days. In the first stage, the rostral and caudal limits of the primary somesthetic cortex (SL) were identified by a brief multi-unit electrophysiological recording. Also in this stage, retrograde neurotracer injections were carried out in Sl and the caudal somatosensory area (SC). In the second stage, a complete mapping of the injected areas was carried out, with a greater density of penetrations, in order to identify the boundaries of SI and its topographical organisation in greater detail. After this second mapping, the animals were perfused and the brains were subjected to histological processing for later analysis. Additional animals that had not been electrophysiologically recorded or injected with neurotracers were processed to obtain histological material to characterise the architecture of the thalamus and cortex.

These procedures are detailed in the following sections.

3.3 Pre-surgical preparation and anaesthesia

On the day before the first surgical stage of the experiment, the animal was treated with 0.5 ml of a combination of benzylpenicillin, dihydrostreptocin base and streptomycin base intramuscularly (LM.) (FORT DODGE Small Veterinary Pentabiotic, 0.58 mg/ml, LM.) as a prophylactic procedure for possible infections during the survival period. In addition, 0.3 ml of vitamin K (Kanakion, Roche, 10 mg/ml, LM.) was also administered intramuscularly as a prophylaxis for haemorrhage. Surgical instruments were sterilised with Germekil (Johnson Wax Professional) or autoclaved.

In order to manipulate the animal so that it could be prepared for surgery, a single anaesthetic dose of 60 mg/kg of sodium thiopental (Thionembutal 0.5g, Abbott) was injected intraperitoneally. This anaesthetic protocol was developed during this study in order to obtain an adequate level of anaesthesia to perform the surgery, while still allowing the identification of cellular responses in the electrophysiological recordings and guaranteeing the animal's recovery during the survival period. The dose used allowed for an anaesthetic level ideal for surgical procedures for approximately 4 hours. It was not possible to recover the animals after administering total doses of more than 60 mg/kg per experiment.

In animals 0409 and 0602, we used a different anaesthetic protocol to the one described above. In these two cases, the anaesthetic level was achieved with a dose of 2 ml/kg of a combination of 9 mg/ml alfaxolone and 3 mg/ml alfadolone (Saffan, Schering-Plough, LM.) injected intramuscularly into the hind paw. This anaesthetic is specifically for veterinary use (cats) and its application in opossums provides excellent results (OLIVEIRA, 2001). However, its manufacture has recently been discontinued. In these cases, in order to administer the maintenance dose of anaesthetic, a superficial vein on the side of the animal's tail was incised after a brief dissection of the skin. An intravenous cannula connected to a continuous infusion pump (Harvard Apparatus) adjusted to a flow of 1.0 ml/kg/h was inserted into this area.

In order to reduce mucus production in the airways, 0.1 ml of atropine sulphate (Ariston; 0.25 mg/ml; LM.) was injected immediately after induction of anaesthesia. Similarly, cerebral oedema during surgery was prevented or minimised with 0.5 ml of dexamethasone (Decadron, Prodome, 4 mg/ml).

Once anaesthesia had been induced, the animal was trichotomised on the hind leg, chest and scalp. The animal was then connected to an electrocardiographic recording system by inserting two syringe needles into the thorax to monitor the heart rate. An endotracheal tube filled with anaesthetic gel (Lidocaine 2%, União Química) was inserted into the trachea using a paediatric laryngoscope, in case artificial ventilation was required. The animal's internal temperature was controlled using a rectal thermoregulator (Frederick Haer Inc.) and maintained at 34.5 °C by an electric blanket.

In the second stage of the experiment, around 10 days later, the animal was anaesthetised with an initial dose of 1.25 g/kg urethane (Sigma-Aldrich) via intraperitoneal (I.P.). This anaesthetic was maintained until the end of the recording session, with up to two additional maintenance doses of 0.42 g/kg I.P. applied when the animal reached a superficial level of anaesthesia. Maintenance doses were usually administered two and four hours after the initial dose, after which the animal usually maintained the anaesthetic level until the end of the experiment.

3.4 Surgical procedures

Initially, the animal was positioned in a Horsley-Clarke stereotactic system (La Precision Cinematographique), duly adapted for the species (OSWALDO CRUZ and ROCHA-MIRANDA, 1968). The craniotomy was preceded by an incision in the skin and removal of the temporal muscle using an electrocautery (Medicir, Medical-cirurgia-Ltda). In all cases, the incision was made on the dorsolateral surface of the animal's left hemisphere.

The skull was drilled using a dental surgical drill (Promeco Ltda) and a surgical stereomicroscope (DF-Vasconcelos, n° 1189). During this procedure, a sterile solution of 0.9% sodium chloride was applied to the skull to dissipate the heat induced by the friction of the surgical drill with the bone. The opening window, of approximately 15 mm, sought to expose the orbital fissure and the entire parietal cortex, and was initially estimated using the stereotactic coordinates described in the atlas of the opossum brain (OSWALDO CRUZ and ROCHA-MIRANDA, 1968). On the medial-lateral axis, the opening extended from an area close to the midline to the most lateral region visually exposed by the separation of the temporal muscle. Finally, the dura mater was carefully cut and moved away from the bony edges so that the surface of the cortex could be visualised.

The cortical surface with its vascular pattern was photographed using a digital camera (Sony, Cyber-shot 3.3 mega pixels) positioned over the surgical microscope's eyepiece. The contrast and brightness of the image obtained were adjusted using an image editing programme (Adobe Photoshop 7.0, Adobe Systems Inc.). The vascular pattern in the printed photograph served as a reference for positioning the sites of penetration of the electrode and neurotracer injections.

After surgical exposure of the cortex, sterile 0.9% sodium chloride solution was dripped directly onto the cortex to prevent it from drying out. When bleeding from the bone tissue occurred, the bone edges were covered with a specific wax for bone haemostasis (Bone Wax W-31, Ethicon).

At the end of the experimental procedures, the cortex was protected with a sterile contact lens (Astralens, Methafilcon 55% H2O) cut to fit over the exposed cortical surface. Next, the edges of the dura mater were repositioned over the surface of the contact lens and the space generated by the bone opening was filled with gel foam (Pharmacia & Upjohn Company). A mass of acrylic material was moulded and positioned over the skull opening window to cover the exposed cortex region, thus avoiding possible mechanical damage. The muscle and skin were then sutured back into their original position. At the end of the experiment, a transdermal fentanyl patch was attached to the animal's hind paw to prevent post-surgical pain.

The same pre-surgical care as described was maintained in the terminal experiment (second surgery), with the exception of the sterilisation procedures for the material, which were now unnecessary. In the second stage, the animal was anaesthetised and repositioned in the stereotactic system. The soft tissues were removed or removed with the electrocautery; and the acrylic cover was removed to re-expose the anterior parietal cortex and the orbital fissure. When necessary, the craniotomy was enlarged with a punch to expose more of the cortical tissue to be recorded.

3. 5 Electrophysiological recording

Neuronal activity was recorded using tungsten microelectrodes (Frederick Haer Inc. and Micro Probe Inc.) coated with varnish and with an exposed conical tip with a diameter of less than 1 *pira*. The impedance of the electrodes ranged from 800 KOhms to 1.2 Mohms, according to the manufacturer. The choice of electrode was made in favour of those that allowed for multi-unit recording. The electrode was attached to an acrylic holder supported by a mechanical micromanipulator. The manipulator was positioned on the rail of the stereotactic apparatus to allow penetration perpendicular to the cortical surface.

The electrical signal was amplified using a high-impedance preamplifier (PS11J, Grass Instruments) next to the electrode and a high-gain amplifier (RPS1070, Grass Instruments). The signal was then sent to an oscilloscope (2120, BK Precision, Model) and an audio monitor. The electrical signal was then converted into a sound signal whose intensity varied according to the greater or lesser activity of the region of cortical tissue being recorded. Under these recording conditions, the signal obtained consists of the superposition of the electrical activity produced by cell bodies and fibres located close to the microelectrode.

In the first experimental stage, electrophysiological recording was only aimed at grossly identifying the representation of the SI forepaw and the borders between the Sl, SR and SC areas. Complete mapping was not carried out in the first surgery in order to reduce surgical time and facilitate the animal's recovery during the survival period. In the second stage, the same procedures and electrophysiological criteria as described were maintained, with the exception of the recording session, which lasted more than 10 hours.

The first penetration site chosen was located around 2.5 mm from the orbital fissure, aiming for the centre of Sl. When the responses to mechanical stimulation showed characteristics of Sl (see description below), the penetrations were continued in a row at average intervals of 500 *pira* between each site, with the recording sites oriented parallel to the midline.

The types of somatosensory stimuli tested in the mapping included: 1) light touch on the skin or manipulation of the hairs or vibrissae; 2) deep touch and pressure of the skin; and 3) passive movement of the joints (proprioception). Stimuli on the skin were generally applied with a wooden stick. During deep skin stimulation, care was taken to minimise joint movement. With the exception of light superficial touching of the skin, all responses to other stimuli were considered to be responses to deep stimuli, including proprioception stimuli.

The electrophysiological criteria for estimating the rostral and caudal limits from S1 included: 1) a decrease or absence of the response to the somesthetic stimulus and 2) a change in the somesthetic modality of the evoked response (from superficial to deep cutaneous, for example) (BECK et al., 1996). The intensity of the neuronal response to the somesthetic stimulus was assessed audibly, without a final assessment of the response peaks. Topographical inversion was not used as there were no responses in the areas immediately rostral and caudal to S1.

3.6 Injection of Neurotracers

Injections were made with the fluorescent neurotracers fluoro-ruby (FR, conjugated to 3 kD dextran, Molecular Probes, Inc.), fluoro-emerald (FE, conjugated to 3 kD dextran, Molecular Probes, Inc.) and diamidino yellow crystal (DY, Sigma). All three tracers chosen were considered retrograde. Retrograde tracers are preferentially taken up by axons and transported to the cell body that originated the projection (VERCELLI et al., 2000). Injecting this type of tracer makes it possible to identify the location of neurons that project to the injected area.

These tracers have a lower scattering than the HRP (REINER et al., 2000; VERCELLI et al., 2000) used by Donoghue and Ebner (1981) to analyse the thalamocortical connections of S1. Another advantage of fluorescent neurotracers is that tracers sensitive to different wavelengths can be injected into the same animal to simultaneously trace the connections of different areas (VERCELLI et al., 2000).

Fluorescent tracers are chemically associated with dextrans, which are hydrophilic polysaccharides conjugated to distinct fluorescent or biotinylated molecules (VERCELLI et al., 2000). Fluorescent dextrans differ from one another in their molecular weight and the fluorochrome to which they are coupled (VERCELLI et al., 2000). In the case of fhioro-ruby, the dextran is conjugated with rhodamine and in fluoro-emerald with fluorescein. Fluorescent dextran differs in molecular weight (VERCELLI et al., 2000). For this reason, we used the lower molecular weight 3kD tracers (3 kD), which are considered ideal for retrograde labelling analysis.

The injections were carried out in Sl and SC with 1 and 10 *jA* Hamilton syringes. Different injections within Sl were carried out in order to identify possible differences in the connection pattern. Each injection consisted of a total volume of 0.4 to 1 *JA*. The syringe was positioned at three different depths in the cortex, ranging from 500 *fim* to 1.5 mm from the cortical surface (Table 4 - in results). At each of the three chosen depths, the syringe remained for 10 to 20 minutes per injection.

The expected survival time for the fluorescent neurotracer to be transported retrogradely was 10 to 12 days. Only then was the terminal experiment carried out for detailed mapping of Sl, SR and SC.

3.7 Perfusion, dissection and flattening

At the end of the electrophysiological recording session, a lethal dose of 10 ml of pentobarbital (30 mg/ml) was administered via LP. After checking for the absence of painful reflexes, perfusion procedures were started by performing a thoracotomy. Next, the pericardium was sectioned and the heart muscle exposed for the application of an anticoagulant - heparin sodium (5,000 Ul/ml Liquemine, Roche). The aorta was ligated above its origin in the left ventricle. The artery was then perforated, downstream of the node, to insert a cannula which, once correctly positioned, was tied into the aorta itself. Through this cannula, perfusion solutions were introduced into the vascular system that irrigates the brain. Finally, the right atrium was also perforated by a small incision, in order to drain the blood and the solutions that returned through the venous system.

The perfusion solutions consisted of: 1) 300 ml of 0.9 % sodium chloride solution (saline solution); 2) 300 ml of 4 % paraformaldehyde fixative solution in 0.1 M phosphate buffer; 3) 300 ml of 10 % sucrose cryoprotectant solution; and 4) 300 ml of 30 % sucrose, both in 0.1 M phosphate buffer.

At the end of the perfusion, the brain was removed from the cranial box. Next, the left telencephalon (injected hemisphere) was carefully separated from the white matter and other subcortical structures to be flattened. During flattening, the cortex was compressed between two glass slides, spaced apart by a small piece of histological slide so that there was no excessive deformation during the flattening process. The diencephalon became a separate block of nerve tissue to be cut coronally.

The material was then placed in a post-fixation solution (4% paraformaldehyde plus 30% sucrose) for a period of 1 day, with the exception of the right hemisphere which, in some cases, was not processed in the neurotracer injection experiments.

The same perfusion protocol described above was used in the architectural analysis experiments. In these cases, during dissection, the hemispheres were separated and the right telencephalon was carefully separated from the diencephalon.

3.8 Cryomicrotomy

For cryotomy, the previously flattened left isocortex and diencephalon were frozen in two different blocks. They were frozen by immersing them in Tissue-teck OCT (Sakura) wrapped in dry ice at -70°C.

The blocks were then placed in a cryostat (Leica CM1850) for about an hour to reach a temperature of -18°C, ideal for microtomy.

3.8.1 Isocortex

The cutting plane of the flattened hemispheres was tangential to the cortical surface, using a thickness of 40 *pira.* The resulting sections were individually immersed in separate containers containing 0.1 M phosphate buffer solution (pH = 7.2-7.4). Alternating sections were taken in separate series for fluorescence analysis and myelin processing (GALLYAS, 1979), the latter being stored for at least 30 days in 10% formalin before being submitted for histological processing. Occasionally, in cases of architectural analysis not subject to electrophysiological recording and injection, the right telencephalon was sectioned in the parasagittal plane for histological analysis of the anterior parietal cortex.

3.8.2 Diencephalon

The 40 or 60 pm coronal sections from the diencephalon were taken in at least two different alternating series, usually one for fluorescence analysis and the other for Nissl staining. In case 0629, a third series was selected for the cytochrome oxidase reaction. In cases 0630 and 0631, we also carried out immunohistochemistry for calbindin and parvalbumin. For Nissl staining, the sections were preserved in containers containing 10% formalin. For the fluorescence series, cytochrome oxidase and other reactions, the sections were collected in containers containing 0.1 M phosphate buffer.

3.9 Histological processing

3.9.1 Cuts for fluorescence microscopy analysis

The sections reserved for fluorescence microscopy were mounted on gelatinised histological slides within 1 day of microtomy. After the sections had dried on the slides, 10% glycerol (VETEC) was dripped onto the slide and the sections were covered with a coverslip and sealed with nail polish. The material was kept refrigerated at 4° C.

3.9.2 Nissl staining to analyse cytoarchitecture

The sections destined for Nissl staining were kept for at least seven days in a 10% formalin solution. After this minimum period of fixation, the sections were mounted on gelatinised histological slides and left to dry for at least 24 hours.

The sections were then dehydrated in ethanol solutions at the following concentrations: 75 %, 95 % and 100 %. The slides remained in each solution for 5 minutes and were then immersed for 10 minutes in a degreasing solution of 100 % ethanol plus chloroform in a 1:1 ratio. The sections were then rehydrated in the ethanol solutions in a concentration sequence of 100 %, 95 % and 75 %, for five minutes each. Then, for one minute, all the sections were washed in distilled water and immersed in cresyl violet solution at 37-40° C for 1 to 2 minutes. Again, the slides were rinsed in distilled water until all the excess cresyl violet was washed off. The sections were then passed through 95% ethanol with 5 drops of acetic acid for 3 to 6 minutes in order to diaphanise the cresyl violet stain. Finally, the mounted sections were passed through a sequence of solutions of 95 % ethanol, 100 % ethanol, 100 % ethanol plus butyl acid (1:1 ratio), for 3 minutes each. Before being covered with Entellan (Merk) as a mounting medium, the slides were dehydrated in xylene for 5 minutes.

3.9.3 Silver impregnation of myelin and myeloarchitecture

The granular layer (IV) of primary sensory cortical areas is generally more myelinated than those of second-order and/or associative areas. The protocol used (GALLYAS, 1979 modified by JAIN et al., 1988) aims to reveal this myelination through silver impregnation and subsequent revelation, creating a contrast between cortical areas with different intensities of myelination.

The histological sections separated for the Gallyas protocol were kept in 10% formalin for at least 30 days. Initially, the sections were washed in distilled water for five minutes and then incubated in a 2:1 solution of pyridine/acetic anhydride for thirty minutes. The sections were then washed three times for three minutes in 0.5% acetic acid. The sections were then placed in ammoniacal silver and kept there for one hour. This solution consists of distilled water, silver nitrate, ammonium nitrate and sodium hydroxide. Again, the sections were washed in 0.5% acetic acid. In the final stage, the sections were then placed in: 1) a developer solution containing sodium carbonate, silver nitrate, ammonium nitrate, tungstosilicic acid, distilled water and 35% formalin for 5 to 10 minutes; 2) distilled water to wash the sections; 3) a potassium ferricyanide (KFeCN) "bleaching" solution and 4) a sodium thiosulphate (NaThioS) fixative solution. After five minutes in the fixative, the sections were washed three times in distilled water and then mounted on a gelatinised slide in a medium of gelatine and 95% alcohol.

3.9.4 Histochemistry for cytochrome oxidase and NADPH-diaphorase

The protocols used below are based on the concept that the presence of the enzymes cytochrome oxidase and NADPH-diaphorase in tissue can be revealed by the conversion of a substrate applied to the tissue, the chromogen, which results in a coloured product at the end of the reaction (HOROBIN, 1982).

3.9.4.1 Cytochrome oxidase (WONG-RILEY, 1979)

The sections separated for the cytochrome oxidase reaction were incubated "ovemight" in a solution containing: 50 ml of 0.1 M phosphate buffer; 5g of sucrose; 25 mg of cytochrome C; 13 mg of diaminobenzidine (DAB, Sigma) and 19 mg of catalase. The sections were protected from light and initially monitored every 1 hour to check the speed of the reaction. The following day, if the sections had not reached the appropriate reaction intensity, a new solution was prepared and the sections were re-incubated.

3.9.42 NADPH-diaphorase (SCHERER-SINGLER et al., 1983)

The sections to be reacted for NADPH-diaphorase were collected after cryotomy in 0.1 M phosphate buffer. Before the reaction, all the sections were washed three times for 10 minutes in 0.1 M tris buffer.

After washing, the sections were incubated in an incubation solution. To prepare this solution, 50 ml of 0.05 M Tris solution (pH uncorrected) was first mixed with 300 mg of D-L-Malic Acid (0.4 M). The pH of the resulting solution was then adjusted to 8.0 with 5N NaOH. After adjusting the pH, the following was added to the solution: 1) 15 mg nitro blue tetrazolium (SIGMA, catalogue no. N-6876); 2) 0.5 ml dimethyl sulphoxide (DMSO, SIGMA, catalogue no. D-5879); 3) 0.5 ml Triton X-100; 4) 18 mg manganese chloride, 50 mg beta-NADPH (SIGMA, catalogue no. N-0505).

The sections remained in the solution under constant agitation, protected from light, and monitored every 1 hour to check the speed of the reaction. If the sections didn't react within the first 6 hours, they remained incubated "overnight" and were removed from the incubation solution the following day.

3.9.43 Immunohistochemistry for calbindin and parvalbumin

Cells of certain biochemical systems have specific interactions between protein regions with particular chemical groups (HOROBIN, 1982). Taking advantage of this principle, immunohistochemical techniques preferentially use antigen-antibody binding to "mark" certain proteins. Calbindin and parvalbumin are proteins that can bind to specific antibodies revealed later in the reaction.

All the sections separated for the calbindin and parvalbumin series were washed in a 1:10 dilution of phosphate-buffered saline (PBSB) twice for 10 minutes and once for 60 minutes before the reaction began.

The sections from each series were then incubated in the primary antibody for approximately 22 hours under constant agitation. The primary antibodies used were from the Vector kit, horse serum. The dilution of the primary antibody for calbindin was 1:2500 and 1:3000 for parvalbumin. The sections were then washed in PBSB three times for 10 minutes and then incubated in the secondary antibody (Vector PK4002 kit, 1:200) for 1 hour. Before the next step of revealing the labelled cells, the sections were again washed three times for 10 minutes. In preparation for development, the sections were incubated in the solution resulting from the ABC kit (Vector) for 1 hour. This solution contained 15 ml of diluent, three drops of reagent "A" and three drops of reagent "B" (ABC Standard kit from Vector). After preparation, it was left to stand for around 20 minutes to form the avidin-biotin complex that characterises the resulting solution "C". After incubation in solution "C", the sections were washed three times for 10 minutes. Finally, the sections were revealed in a 0.05% DAB solution for approximately three minutes. This solution consisted of DAB (Sigma-Aldrich), PBSB and 3% hydrogen peroxide.

3.10 Analysing and documenting the results

3.10.1 "Plotting of fluorescently labelled cells

The sections mounted for fluorescence were analysed on a Leitz microscope (Orthoplan 2) equipped with an epifluorescence source and filters suitable for observing the cells retrogradely marked by the fluorescent tracers.

The sections were drawn with the aid of an x and y axis encoder on the microscope platen, connected to a computer equipped with the cell plotting programme MD-Plot version 3.3 (Minnesota Datametrics Corporation). The position of the cell bodies of the retrogradely marked cells in the thalamus and cortex was digitised. In addition, the outline of the injection sites, blood vessels, lesions and the outline of the histological sections where the marked cells were found were drawn on this system. The drawings were made with lOx (thalamus) and 25x (flattened cortex) lenses.

The percentage of retrogradely labelled cells in the thalamus was calculated in the most complete cases. The final count was made on all the plotted sections, including all the identified thalamic nuclei. The final result was illustrated by showing the percentage of labelled cells in the somatosensory nucleus (VB) and the motor nuclei (VL, VA and VM). The rest of the thalamic nuclei were considered together.

3.10.2 Delimitation of the thalamic nuclei

The thalamic nuclei were identified in sections stained using the Nissl method and/or cytochrome oxidase. Delineation of the nuclei began by adjusting the gain of the sections to allow alignment with the drawings resulting from the fluorescence plot. Drawings of the thalamic architecture were made using a microprojector for histological slides (Ken-Vision) or an aristophoto (Eleitz Wetzlar). The overlap between the Nissl stained section and the adjacent section used for fluorescence analysis was based on the contour of the section and the position of the blood vessels present in both sections. Once this overlay had been made, the boundaries of the nuclei were drawn over the plot of the labelled cells. The drawing resulting from this overlay was scanned and transferred to the Canvas X graphics programme (ACD-Systems) for final finishing.

3.10.3 Delimitation of cortical areas

The cortical areas were identified by delimiting the edges shown in the myelin marking method.

Delineation of the cortical areas was made possible by photomicrography of the best myelin-stained section that allowed visualisation of the cortical areas (2.5x). The outlines of all the areas identified were made in the Canvas X programme, which provided information on the size of the area occupied by each cortical area and by the animal's neocortex. In some cases, more than one section was drawn on the aristophoto and the result of the superimposition showing the architectural boundaries was used for analysis.

Once the electrophysiological and myeloarchitectural borders of the Sl, SR and SC areas had been identified, all the fluorescence sections plotted on the MD-Plot were superimposed on the electrophysiological and architectural maps to unite the retrograde fluorescent marking information with the borders of the areas. This overlay was carried out in the Canvas X programme, based on the contour of the sections, the position of the blood vessels and the injection sites.

The measurement of the area occupied in the neocortex by Sl, VI, Al and the frontal cortex was obtained automatically by creating a contour around these areas - a feature available in the Canvas X programme.

3.10.4 Photomicrograph

The sections used to analyse the architecture of the cortex and thalamus were photographed using a camera attached to a Leitz microscope (Orthoplan 2) with 2.5x, 5x and 6.3x objectives. Image adjustments were made in the Abobe Photoshop programme version 5.5 before the figures were assembled. The resulting photograph was transferred to the Canvas X graphics programme for final processing.

CHAPTER 4

RESULTS

In the following results, we aim to describe the thalamo-cortical and cortico-cortical projections for somatosensory areas Sl and SC in the opossum *Didelphis aurita.* To do this, we used more than one criterion for delimiting cortical borders, the results of which will be described below. Firstly, we will describe the architectural subdivisions of the neocortex of the opossum *Didelphis aurita* using the Gallyass technique (myeloarchitecture). In order to define the region where the neurotracers were injected, we also describe the electrophysiological responses that characterise the Sl, SR and SC areas. The mapping data was corroborated by the overlap with the myeloarchitecture obtained.

Next, we will re-analyse the architectural subdivisions of the motor and somatosensory thalamus of this same species, comparing the marking obtained by the classic Nissl method, with the cytochrome-oxidase, NADPH-diaphorase, calbindin and parvalbumin techniques, used for the first time in the thalamus of this species, to ascertain the limits of the thalamic nuclei that project to the somesthetic cortex.

Finally, after recognising the cortical areas and thalamic nuclei to be studied, we will describe the pattern of thalamic and cortical projections through neurotracer injections in the Sl and SC areas of the neocortex of the opossum *Didelphis aurita.* Injections of neurotracers into the SR of this species were carried out previously by Santiago Martinich (1996) in the post-orbital region and also by Donoghue and Ebner (1981). The SC area was emphasised in our study in order to verify the possible homology of this region with other somatosensory areas caudal to the Sl characteristic of the sensory-motor system of mammals with complex brains, such as areas 1,2 or the posterior parietal cortex (KAAS, 2004).

4.1 Anatomical and functional characterisation of the somesthetic areas of *Didelphis aurita*

4.1.1. Mieloarchitecture

The myeloarchitecture of the cortex of the opossum *Didelphis aurita* was analysed using tangential sections, which included cuts from the piai surface to the white matter, processed using the technique of Gallyas (1979) modified by Jain et al. (1988). The sections that best allowed the edges of the sensory areas to be identified were chosen for photographic illustration.

The primary sensory areas stand out for being more densely marked with myelin than the rest of the neocortex (Figure 11). In the four cases analysed, the primary visual cortex stood out as the most extensive, followed by the primary somatosensory cortex Sl (Figure 12a-d). In general, the auditory cortex tended to occupy a smaller portion of the neocortex, as did the frontal (non-sensory) cortex, which also showed dense myelin labelling (Figure lla-c). The frontal region was intensely marked by the Gallyass technique and seems to include the F4 area mentioned by Martinich et al. (2000).

Of particular interest to our study, the Sl, SR and SC areas could be easily distinguished from each other by myelin labelling. As described above, the Sl area was densely marked (Figure 11). The lighter region immediately rostral to Sl and caudal to the orbital fissure was identified as SR (BECK et al., 1996). The SC area showed intense myelin marking, although it was less densely marked than Sl. An exception was case 0705 (Figure 11b), where the SC area was very lightly marked.

The peristriate cortex in *Didelphis aurita* was identified between VI and the parietal cortex, according to Martinich et al. (2000). Still in the medial region of the peristriate cortex, it was possible to observe a highly myelinated region with a rounded shape (Figure 11a and 11c) similar to those of the

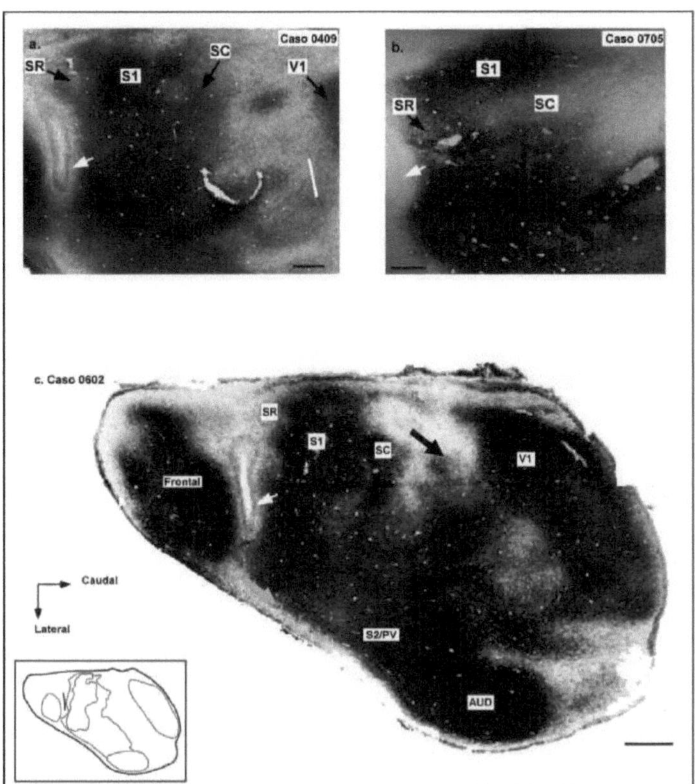

Figura 11. **Photomicrograph of the sensory areas of the opossum *Didelphis aurita* revealed by myeloarchitecture.** (a) and (b) show the Sl, SR and SC areas in greater detail, in two different cases. The central area Sl is the most densely marked for myelin (in black), while areas SR and SC are characterised by slightly less dense marking than Sl. In (c), a third example, at lower magnification, showing VI and the auditory cortex (AUD) which also show intense marking for myelin. The schematic drawing in the bottom left-hand corner of the figure illustrates the subdivisions proposed for case 0602. The white arrow in the three cases points to the orbital fissure. The black and grey arrows represent the location of the peristriate cortex and part of the insula, respectively. Calibration bar in (a) and (b) = 1 mm; in (c) = 2 mm.

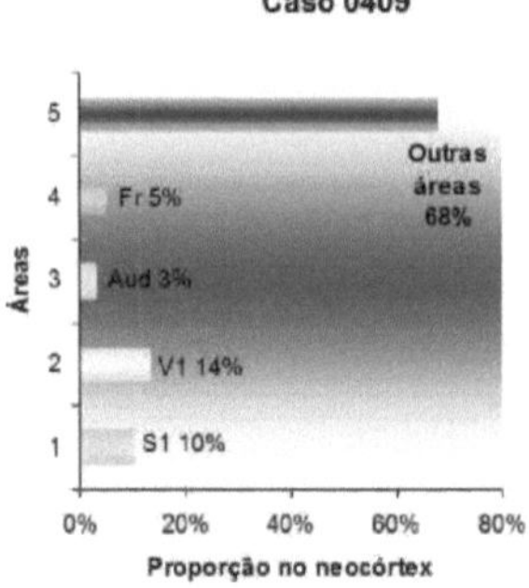

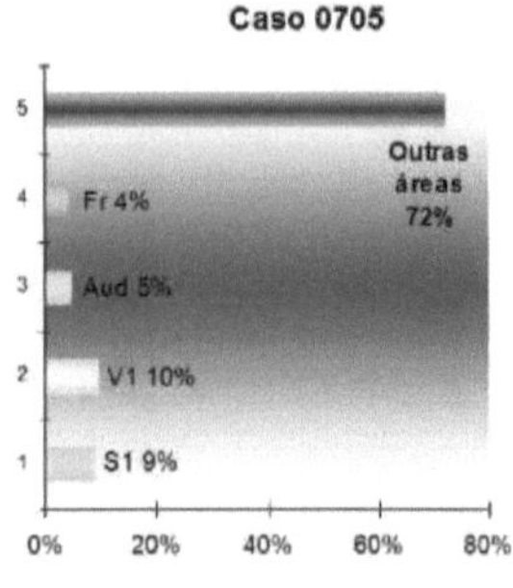

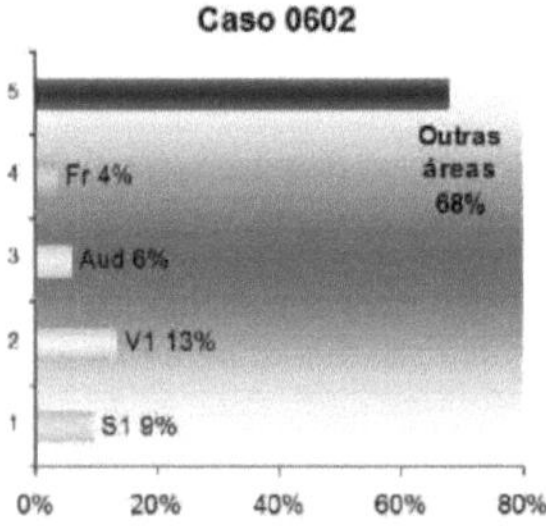

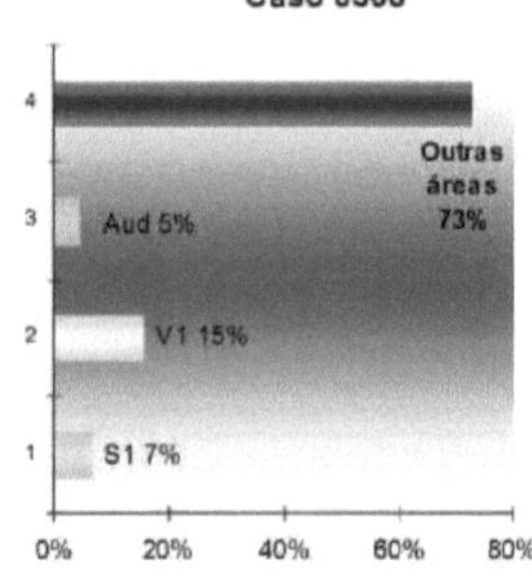

Figura 12.   Proportion occupied by cortical areas in the opossum *Didelphis aurita.* Quantitative analysis of the area occupied by the primary sensory areas and frontal area (Fr) in the opossum neocortex, based on myelin marking. In all the cases illustrated, the primary visual cortex (VI) stood out as the most extensive primary sensory area, followed by the somatosensory cortex (Sl). In case 0506, it was not possible to delimit the Fr area. Aud = auditory cortex.
observed in this species and in *Monoldelphis domestica, using* histochemical techniques (MARTINICH et al., 1990; FRANCA, 1999).

4.1.2 Electrophysiological characteristics of Sl, SR and SC

4.1.2.1 Mapping   the primary somatosensory cortex (SL)

The electrophysiological characteristics of the anterior parietal cortex of the opossum *Didelphis aurita* were studied in five animals (Table 2).

The anterior parietal cortex was mapped through detailed multi-unit electrophysiological recording of the SR, Sl and SC cortices. To compose the electrophysiological map, 62 to 94 penetrations were performed per experiment (see Table 2). After histological processing of the cortical tissue, the electrophysiological map was superimposed on the photomicrograph of the myelin marking pattern, thus enabling the borders of the Sl,

SR and SC areas to be delimited.

Summary of electrophysiological recording results		
	No. of registration sites realised	**Approximate length of Sl (mm)**
Case 0409	-	4,6
Case 0506	62	--
Case 0602	82	3,86
Case 0629	99	1,96
Case 0701	94	-
Case 0705	84	3,2

Table 2. Number of penetrations of each experiment carried out in the parietal cortex to delimit areas Sl, SR and SC by electrophysiological mapping. The drawings resulting from the analysis of myeloarchitecture and electrophysiology made it possible to identify the extension occupied by Sl in the anterior parietal cortex and the estimated distance from this area to the orbital fissure.

According to our results, the Sl area occupies an average length on the rostro-caudal axis of 3.5 mm, with its centre located approximately 3 mm from the orbital fissure.

Within Sl, in all the cases analysed, there was a predominance of recording sites with skin responses (to light or superficial touch, Figure 13). However, some sites only showed a response to deep stimulation of the skin receptors (deep touch), light squeezes or weak responses to superficial stimulation of the skin receptors. This was observed in both the front paw and face representations (Figure 15). Also within Sl, we obtained some recording sites with no response to somesthetic stimulation. In general, the non-responsive sites were retested at a later date to check for possible interference from anaesthesia, which could occasionally be reducing the neuronal response.

The main aim of the electrophysiological mapping was to characterise the area where the neurotracer had been injected. Considering that the face representations in Sl and S2 of the opossum are neighbouring and have a common border (BECK et al., 1996), we decided that the neurotracer injections would be carried out in the anterior paw representation, so that there would be no risk of injecting into the S2 face representation by mistake. As injections into the anterior paw representation of the primary somesthetic cortex were favoured, most of the mappings carried out were limited to detailing mainly this region in Sl.

As previously described (BECK et al., 1996; LENDE, 1963; PUBOLS et al., 1976), we noticed that the representation of the front paw in the primary somatosensory cortex of the opossum is medial to the representation of the face (Figure 14-17). Both in the

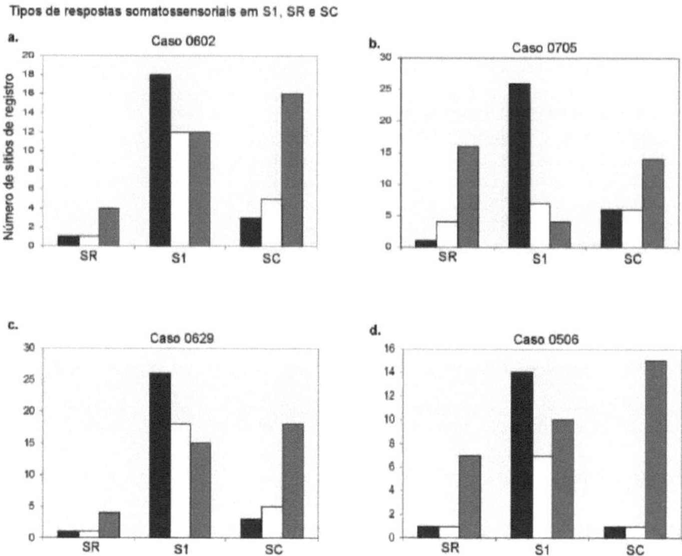

Figura 13.   Quantification of the types of somatosensory responses present in the Sl, SR and SC areas in anaesthetised preparations. In all three areas there are penetration sites responsive to light touch (black bar) and deep somatosensory stimulation (white bar, a-d). The Sl area is characterised by a predominance of responses to light touch (a-d), while the SR and SC areas are distinguished by a predominance of sites unresponsive to somesthetic stimuli (grey bar, a-d). For standardisation purposes, the recording sites quantified refer to the second electrophysiological recording session of each case presented.

both on the face and on the paw, the multi-unit receptor fields tend to be small, representing restricted parts of the body, such as a single phalanx, cushion or vibrissa (Figure 15).

The rostral and caudal borders of Sl were obtained by combining electrophysiology with myeloarchitecture. In general, these borders tend to be approximations. If the electrophysiological border didn't coincide with the myeloarchitectural border (Figure 17), we observed the recording sites close to the architectural border: we considered sites with large receptor fields representing, for example, the entire foreleg or forearm, and/or with a response to deep stimulation, or even non-responsive sites, to be outside of Sl. Recording sites with an intense response to light touch generally surrounded by non-responsive sites were eventually located outside of Sl, depending on their location.

Our results also show that the representations of the distal and proximal segments of the foreleg are orientated along the rostro-caudal axis of S1, with the most distal segments close to the SR area and the proximal segments progressively further back in S1 (Figures 15 and 17). In general, after depicting the phalanges, continuing towards the caudal edge of S1, the cushions are progressively depicted, followed by the wrist (Figure 15 penetrations 1, 2 and 3, respectively; Figure 16 penetrations 1-5). In some cases (Figure 16), the representation of the forearm was identified within S1, caudal to the representation of the wrist (Figure 16b penetration 5).

The arrangement of the five fingers in the primary somatosensory cortex seems to be orientated on the medial-lateral axis. In cases 0629 (Figure 15) and 0705 (Figure 17), it can be seen that the first finger (thumb) is represented lateral to the representation of the other fingers, while the representation of the fifth finger is located more medially in the cortex. In case 0629 (Figure 15), the fifth finger, the ulnar region of the foot, wrist and forearm are found in the same rostro-caudal sequence of penetrations (Figure 15, sites 1 to 4). In the row immediately lateral to this, we found representations of the third finger and the palmar pad immediately proximal to it (Figure 15, sites 5, 6 and 7). In this same row, we observed representations of pad 4 (Figure 15, site 8). In case 0705, fingers 5, 3 and 1 are represented sequentially from medial to lateral (Figure 17, penetration sites 3, 10 and 12, respectively). This organisation resembles the dorsal view of the paw, with the distal phalanges of the fingers pointing towards the rostral pole of the neocortex. In some cases, however, an organised topography of the anterior paw cannot be observed (Figure 15, penetration sites 14 to 21).

Continuing laterally in the cortex, the border between the representation of the front paw and the face in S1 is identified. The border between the face and paw representations is characterised electrophysiologically by penetrations with deep touch responses and larger receptive fields (Figure 14 to 17). The same principle could be extended to the rostral and caudal edges of S1, as shown in Figure 16. The penetrations that characterised these borders also showed responses to deep stimuli and larger receptor fields (Figure 16, sites 1e6).

The topographical representation of the vibrissae in S1 can be assessed in experiment 0629 (Figure 15). The multi-unit receptor fields found could either involve multiple vibrissae from the same row, or represent a single vibrissa. Furthermore, this result suggests that vibrissae from the same column are represented on the rostro-caudal axis of the cortex. As shown in Figure 15g, the second vibrissa (VII) of the lower row is represented in the rostral region of Sl (site 2), while vibrissae VII of the more dorsal rows are represented successively in the caudal regions of Sl (recording sites 33-36, Figure 15).

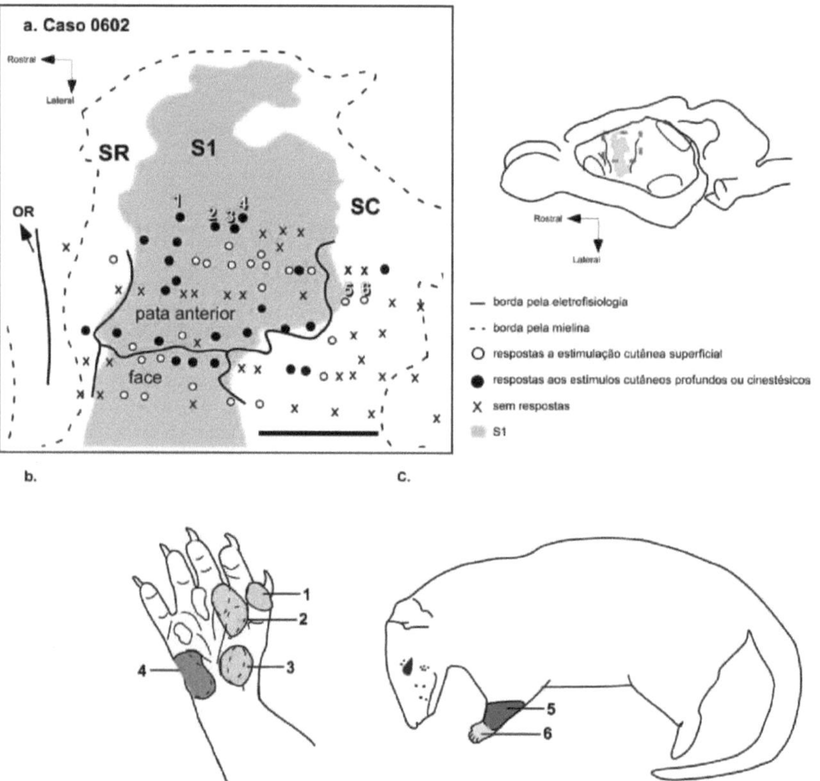

Figure 14. Somatotopic organisation of the anterior paw of SI (experiment 0602). In (a) the green area illustrates the extension of SI in this case by the myeloarchitecture. The continuous and dotted black lines represent the orbital fissure and the architectural border of SR (rostral) and SC (caudal), respectively. The receptor fields corresponding to sites 1 to 4 located in SI (b) are small and represent a single phalanx or cushion. Sites 5 and 6 in SC have larger receptor fields (c). The conventions adopted for this figure will also be used in figures 20 to 22. OR = orbital fissure. Calibration bar = 1mm.

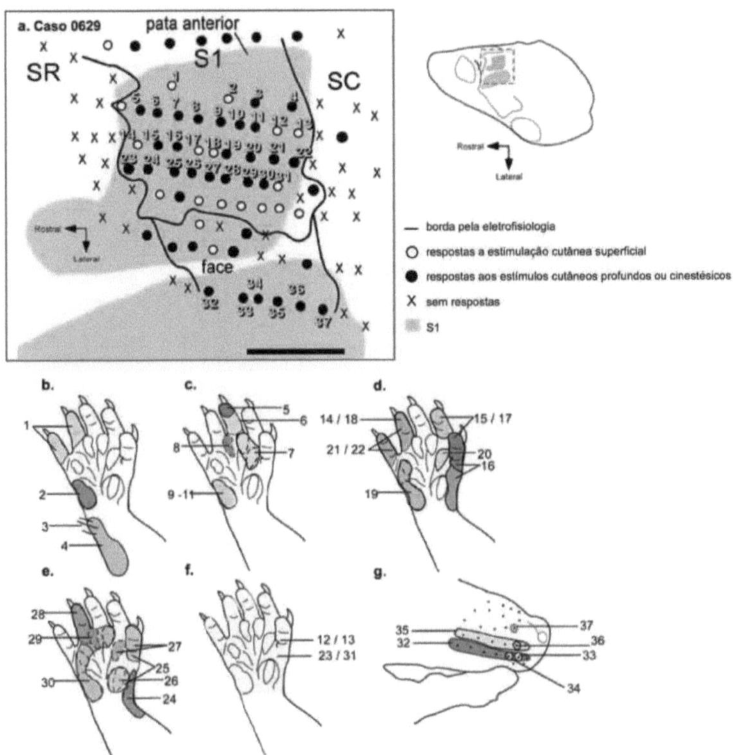

Figura 15. Somatotopic organisation of the anterior paw and SI face (experiment 0629). The receptor fields in (b) corresponding to sites 1 to 4 located in SI (a) represent the ulnar region of the fingers, paw, wrist and forearm. Sites 5 to 7, positioned more laterally in SI (a) have receptor fields for the central region of the hand (c). Sites 32-37 (g) represent a row of penetrations located more laterally in Sl, orientated on the rostro-caudal axis, in the representation of the face. The respective receptor fields progress from the vibrissae located in the lower rows (ventral) to the upper ones (more dorsal). Calibration bar = 1mm.

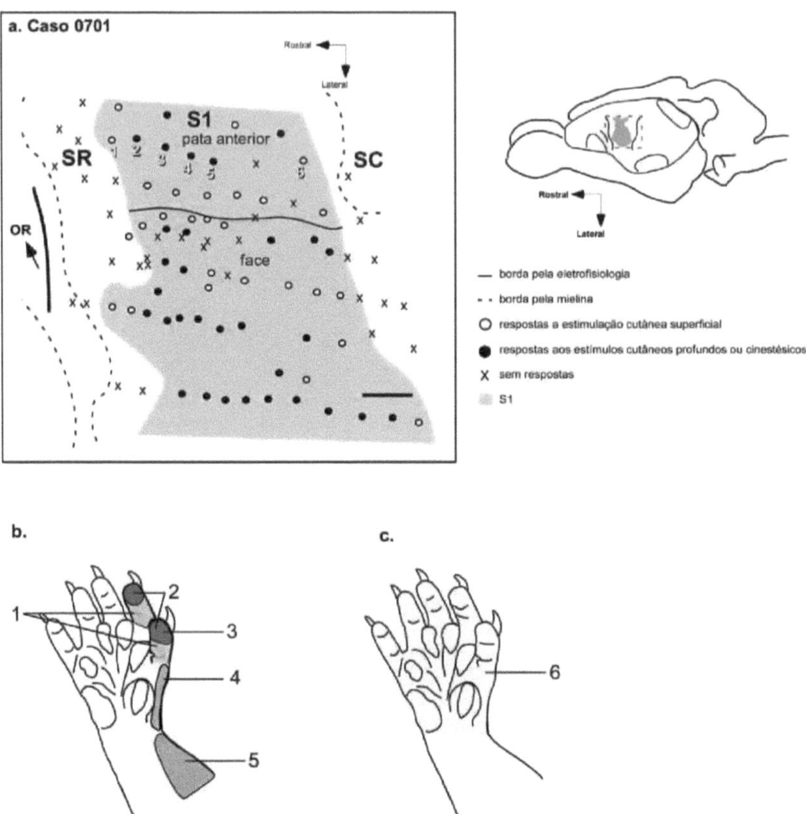

Figura 16. Somatotopic organisation of the SI forepaw (experiment 0701). The sites

1-6 (a) located in SI represent a row of penetrations orientated on the rostro-caudal axis. In (b) the respective

receptor fields progress from the most distal segments of the hand to the proximal ones, such as the forearm.

The edges of SI with SR and SC correspond to penetrations (sites 1 and 6) with a response to deep stimulation

(white circles) and larger receptor fields (c). Calibration bar = 1mm. OR = Orbital fissure

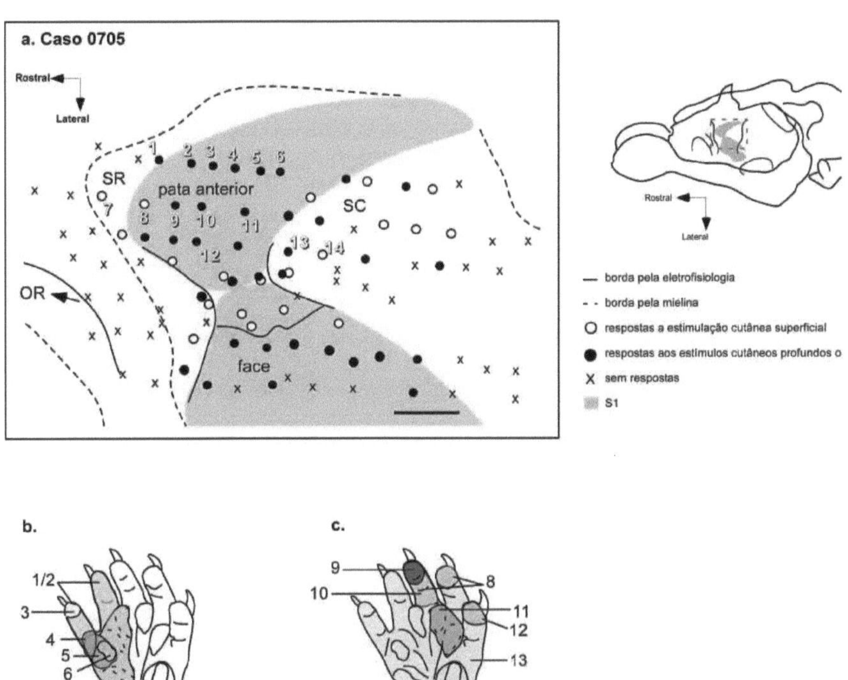

Figure 17. Somatotopic organisation of the anterior paw in SI (experiment 0705). The receptor fields in (b) corresponding to sites 1 to 6 located in SI (a) represent the ulnar region of the fingers and hand. Sites 8 to 11, positioned more laterally (a) have receptor fields for the central region of the hand (c). Penetration site 7 located in SR (a) corresponds to three pads in the ulnar region of the foreleg (b). Sites 13 and 14 (a) located in SC represent the entire foreleg and forearm (c). In both areas, the sites illustrated showed a response only to deep stimulation (white circles). Calibration bar = 1mm. OR = Orbital fissure.

4.1.2.2 Caudal (SC) and rostral (SR) somatosensory cortexes

The caudal somatosensory cortex (SC) corresponds to an area located between SI and the peristriate cortex (MARTINICH et al., 2000), being identified as a region with an intermediate intensity of myelination between SI (more densely myelinated) and the peristriate cortex (little myelinated) (Figure 11).

Figure 13 shows the prevalence of penetration sites without responses in the SC area, which makes it difficult to establish an electrophysiological border with the peristriate cortex, since this region is also

unresponsive in anaesthetised preparations. However, it was possible to record some responses to superficial

and deep tactile stimulation (Figure 13), which reinforces the characterisation of this region as a somesthetic

processing area.

The extent of CS in the rostro-caudal axis, assessed by myeloarchitecture in two different cases, is

quantified in Table 3.

The rostral somesthetic area (SR) comprises the region located between the orbital fissure and the

rostral border of Sl (Figure 11). The SR area is electrophysiologically characterised by the predominance of

penetration sites that are non-responsive to somesthetic stimulation (Figure 13).

	SC extension (mm)^2	Distances from SC centre to crack (mm)
Case 0409	13	5,3
Case 0506	-	-
Case 0602	1,8	5,7
Case 0705	-	-

Table 3: Measurements of the SC area. Area occupied by SC in the anterior parietal cortex and distance
from the centre of this area to the orbital fissure (measured from halfway along the fissure).

Neurons in both SC and SR generally have larger receptor fields than those recorded in SL. Figure 14

(case 0602) illustrates the receptor fields for sites 5 and 6 in SC, which corresponded to the entire foreleg and

forearm, respectively (Figure 14A and C). The same result was obtained in case 0705 (Figure 17). The

recording sites in SC (13 and 14) and SR (7) respectively, had larger receptive fields than those described for

Sl. It is possible that both the SR and SC areas have a complete representation of the body that could not be

observed by our work or by Pamela Beck (BECK et al., 1996), because in addition to the responses found after

stimulation of the anterior paw, responses to stimulation of the face were also obtained in both cortical areas,

but with few penetrations (data not shown).

4.13 Cortico-cortical projections for Sl and SC

Table 4 lists all the neurotracer injections carried out in this study. To summarise, in cases 0602 and

0629 more than one injection was made in different regions of the anterior paw representation in Sl, one more

rostral and the other caudal. In case 0629, two injections were carried out, but we had technical problems

processing the cortex. Therefore, only the thalamocortical projections could be studied in case 0629. All the injections covered the same depth in the cortex, with the exception of case 0705 and case 0602 (injection in

the representation of Sl's face and paw, respectively). We did not observe a spread of the neurotracer to the

white matter in the cases presented. In animals 0705 and 0701, Sl injections were made in both the face and

paw representations, allowing us to compare the connections made by different topographical representation

fields. Figure 18 illustrates injection sites and labelled cells corresponding to FE (A,B), DY (C,D) and FR

(E,F) injections, respectively.

	Area injected	Neurotracer	Distance from the orbital fissure	Spreading area	Survival time	Depth of injection
Case 0409	Edge Sl/SC	FR	4.5 mm	0.0645 mm²	12 days	-
Case 0602	Slpata	FE	1.7 mm	0.0194 mm²	11 days	Surface area up to 1440 µm
	SI paw	DY	3.8 mm	0.129 mm²	11 days	Surface area up to 840 µm
	SC	FR	5.1 mm	0.0258 mm²	11 days	Surface area up to 1440 µm
Case 0629	SI paw	DY	2.5 mm	0.9677 mm²	11 days	-
	SI paw	FE	3.6 mm	0.0258 mm²	11 days	-
	SC	FR	4.6 mm	0.258 mm²	11 days	-
Case 0701	SI paw	DY	2.7 mm	-	12 days	-
	SI face	FE	2.6 mm	-	12 days	-
	SC	FR	3.8 mm	-	12 days	-
Case 0705	SI/SR/SC	DY	1.5 mm	13548 mm²	10 days	Surface area up to 1200 µm
	SI face	FE	2.3 mm	0.0645 mm²	10 days	80 µm to 1200 µm
	SC	FR	4.2 mm	0.1935 mm²	10 days	Surface area up to 1200 µm

Table 4 - Summary of neurotracer injections. Data from all the experiments carried out involving the

injection of neurotracers into the somesthetic cortex. FR = fluoro-emerald; FE = fluoro-ruby; DY = diamidino-

yellow.

4.1.3.1 Primary somatosensory cortex (Sl-pata anterior)

In **_case 0602_** (Figure 19), an injection of FE was made in Sl close to the border with SR and another

injection of DY more caudally. Figure 19b shows the receptor fields corresponding to the injection sites,

corresponding to the representation of the wrist and forearm, respectively. Very few cells were retrogradely

labelled by FE, mostly close to the injection site in the Sl and SR areas (Figure 19c). However, some cells were also visualised in medial portions of Sl. In area S2/PV, very few neurons were marked by FE.

Also in case 0602, the injection of DY into the caudal region of Sl (Figure 19) resulted in labelled cells in the somatosensory areas Sl, SC and SR. The injection site corresponded to the forearm representation of the primary somesthetic cortex (Figure 19b). Similar to the FE injection in this same case, the neurons retrogradely labelled by DY were limited to the area surrounding the injection site (Figure 19c) in both Sl and SC. Marked cells were also identified in SR, medial to the injection site. This injection did not result in labelling in S2/PV.

In ***experiment 0705*** (Figure 20), FE was injected into the representation of the face in Sl. The retrogradely labelled cells were restricted to the somesthetic cortex, with the neurons around the injection site standing out. A group of cells was found in the medial region of the neocortex. These cells extended from the injection site to SC and extended into a zone medial to the somesthetic cortex, which seems to correspond to the PM area previously described in this species (MARTINICH et al., 2000).

The injection of the neurotracer DY into the representation of the anterior paw in case 0705 (Figure 21) resulted in a large spread of the neurotracer, causing intense labelling in the cortex. Unlike the pattern obtained in the injections described above, a large number of labelled cells were obtained in Sl, SR, SC, the peristriate cortex, the auditory cortex and the frontal area (Figure 21).

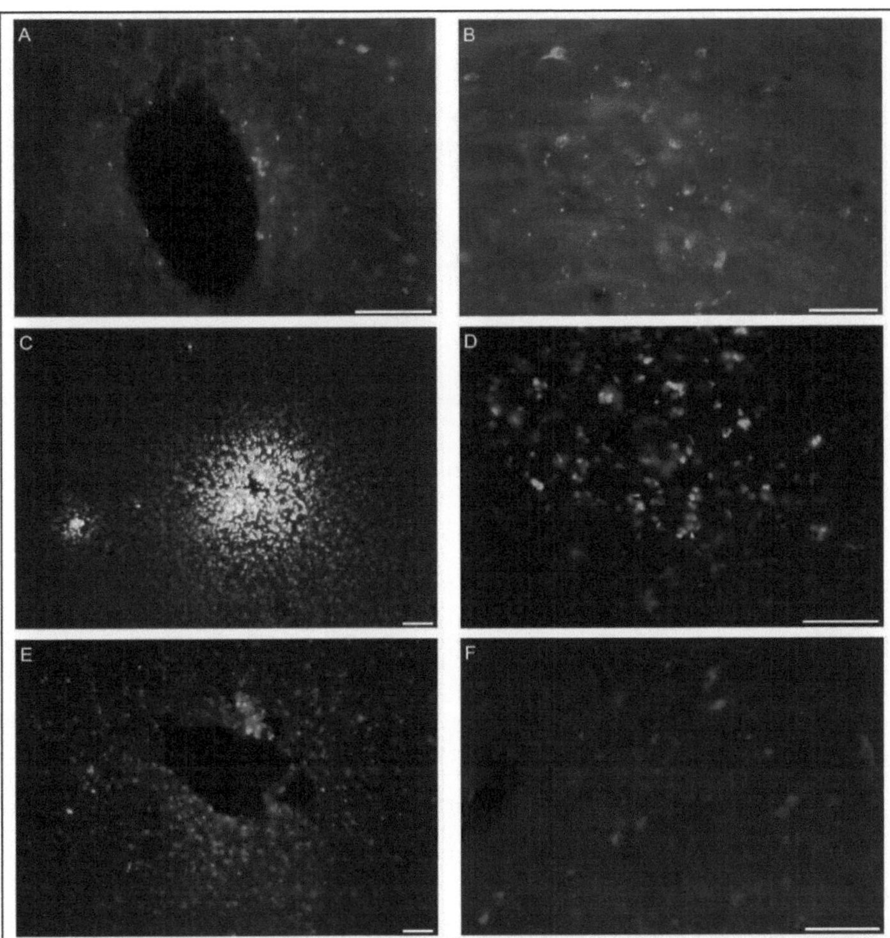

Figura 18.   **Photomicrograph of injection sites and labelled cells.** Note that the fluoro-emerald injection site (a) shows less scattering. In (b) we can see retrogradely labelled cells in the VB nucleus (b). The DY site (c) and the FR site (e) were easily identified. (d) and (f) show cells retrogradely labelled by DY and FR, respectively, in the VB nucleus (case 0629). Calibration bar (A, B, D and F) = 100 μm; (C and E) = 50 μm.

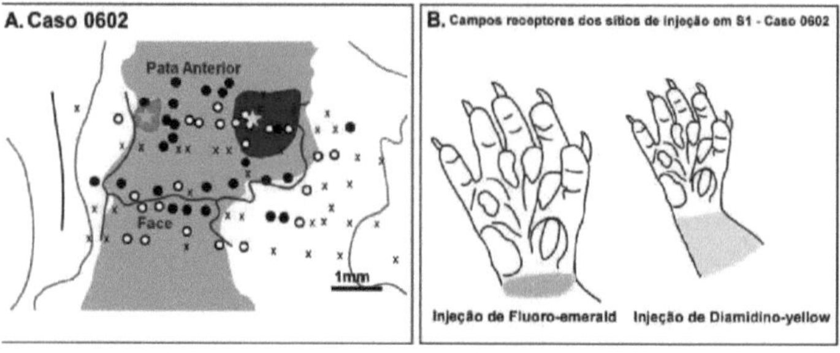

C. Caso 0602

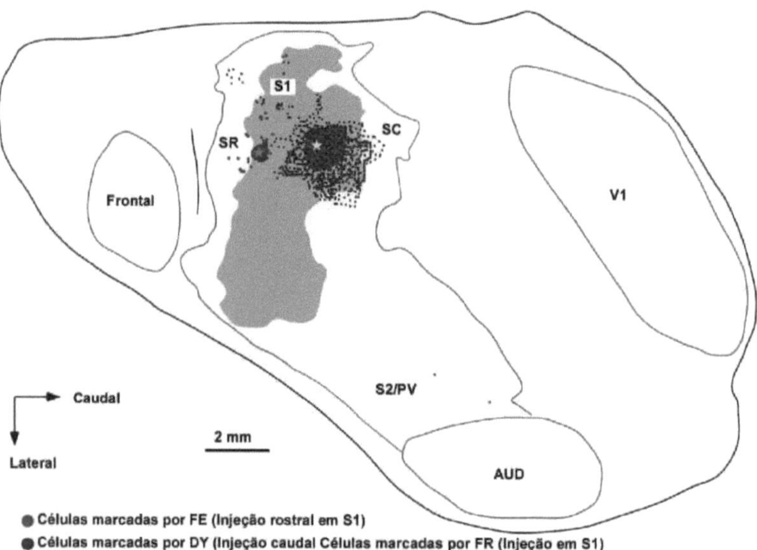

Figura 19. Cortico-cortical projections for SI (Case 0602). Reconstruction of cross-sections of the neocortex. FE and DY injections were performed in the anterior paw representation of SI (a) and (b). Most of the labelled cells were observed within SI, around the injection site (c). Few labelled cells were visualised in SR, SC and S2/PV. Red line = electrophysiological border of S1.

64

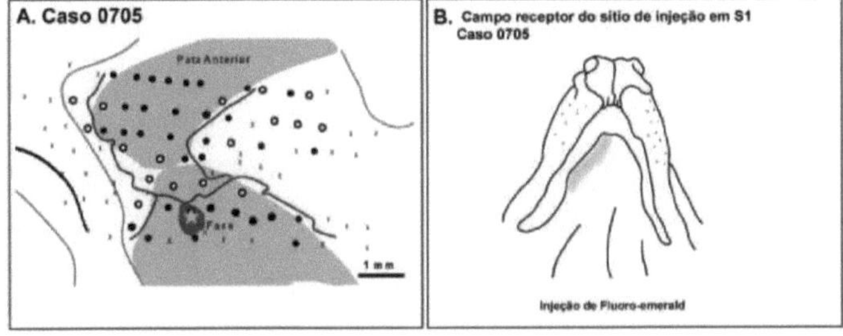

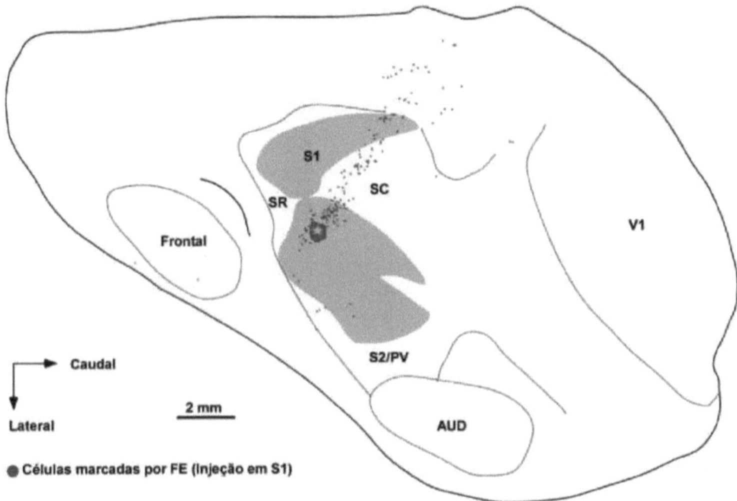

Figura 20.   Cortico-cortical projections for representing the face in Sl (Case 0705).

Cross-sectional reconstruction of the neocortex. In this case, the FE injection was positioned in the

representation of the face of Sl (a) and (b). Most of the labelled cells were observed within Sl, around the

injection site (c). Few labelled cells were visualised in SC, the medial zone of the neocortex and frontal.

65

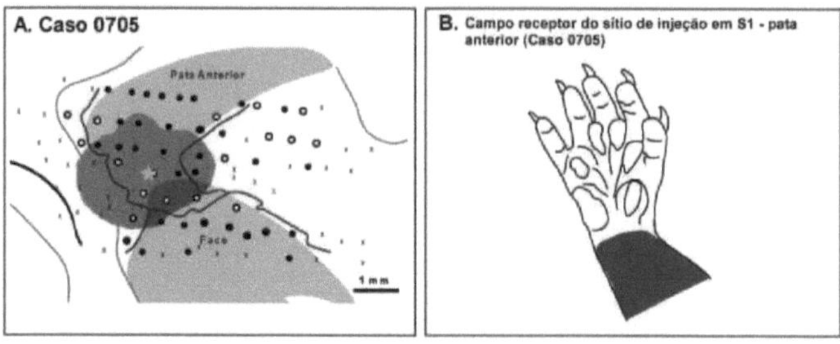

B. Caso 0705

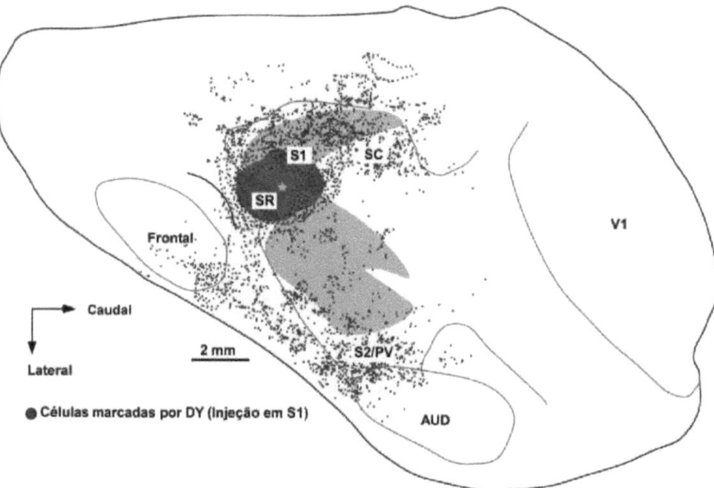

Figura 21. **Cortico-cortical projections for the representation of the anterior paw in area Sl (Case 0705).** Cross-sectional reconstruction of the neocortex. In this case, the DY injection was positioned in the representation of the face of Sl, with spreading of the neurotracer to SR and SC (a). The labelled cells were observed in Sl, SR, SC, S2/PV, auditory cortex, peristriate cortex, frontal cortex and medial zone of the neocortex (c).

4,132 Caudal somatosensory cortex (SC)

Four different injections were made in the caudal somesthetic area in 3 different cases (Cases 0602, 0705, 0409). As with Sl, the position of these injections varied within the anteroposterior axis of SC (Table 4).

In ***case 0602***, FR was injected into a region of the electrophysiological map unresponsive to somesthetic stimulation (Figure 22a), but close to sites responsive to stimulation of the anterior paw (Figure

14, penetrations 5 and 6). The retrogradely labelled cells after this injection are found not only in the somesthetic cortex, but also in the auditory cortex and peristriate visual cortex (Figure 22b). In the somesthetic cortex, a large contingent of cells stood out within SC, in a position medial to the injection site. The cells labelled in Sl and SR showed a similar distribution, with a greater number of labelled cells in the medial cortical regions. However, it was also possible to notice sparse retrogradely labelled neurons along the entire length of Sl and SR. In S2/PV, another group of cells was visualised, close to the auditory cortex, which seems to avoid the representation of the face in these areas *(cf.* BECK et al., 1996).The auditory cortex, in turn, showed a few neurons scattered in its posterior half (Figure 22b). In the region rostral to the auditory cortex, corresponding to the insular cortex (MARTINICH et al., 2000), few labelled cells were found (Figure 22b).In the visual cortex, labelled neurons were only identified in the peristriate cortex. No labelled cells were observed in VI in this case (Figure 22b). A group of labelled cells was found in an area medial to the peristriate cortex. This region seems to correspond to the PM cortex (MARTINICH et al., 2000).In ***case 0705,*** the injection of RF into the SC occurred in a region characterised by penetration sites responsive to deep stimulation of the anterior paw (Figure 23a and b), more specifically the first and second toes.As in case 0602, the injection of RF into the SC marked cells in the somesthetic, visual and auditory cortex (Figure 23c). Again, there was intense labelling of neurons in the region medial to areas Sl, SR and SC. However, in case 0705, the labelling in Sl differed from case 0602 in that neurons were labelled in the region representing the anterior paw. In both cases 0602 and 0705, sparse retrogradely labelled neurons were located in the face representation of Sl. In area S2/PV, cells were observed close to the auditory cortex. Again, we observed neurons in the region lateral to the myelin marking, corresponding to the insula (MARTINICH et al., 2000).

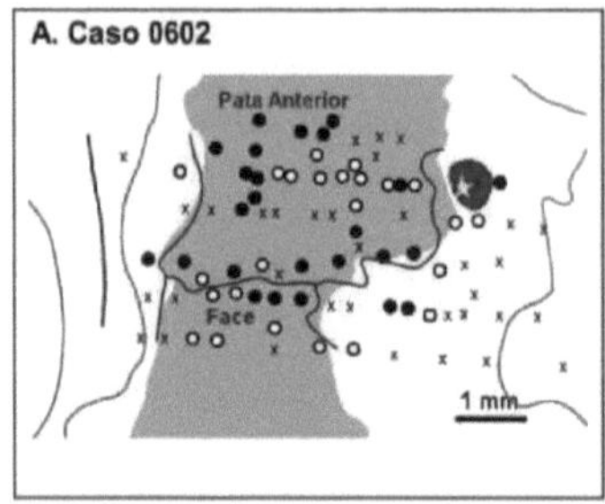

B. Caso 0602

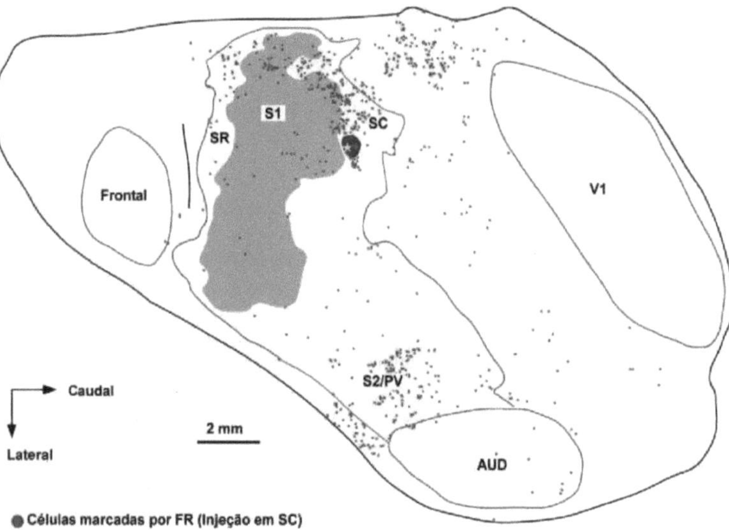

Figura 22.   Cortico-cortical projections for area SC (case 0602). Reconstruction of cross-sections of the neocortex. The FR injection was positioned in SC (a). In the anterior parietal cortex, most of the labelled cells were observed in the medial region of Sl, SR and SC (b). Labelled cells were also observed in S2/PV, auditory cortex, peristriate cortex, frontal cortex and the medial zone of the neocortex (b).

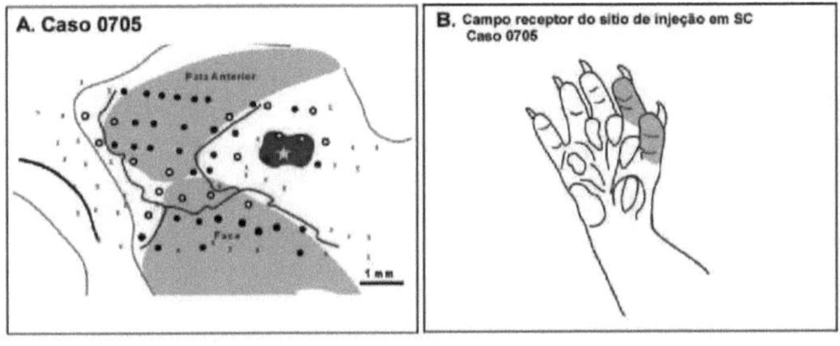

C. 0705

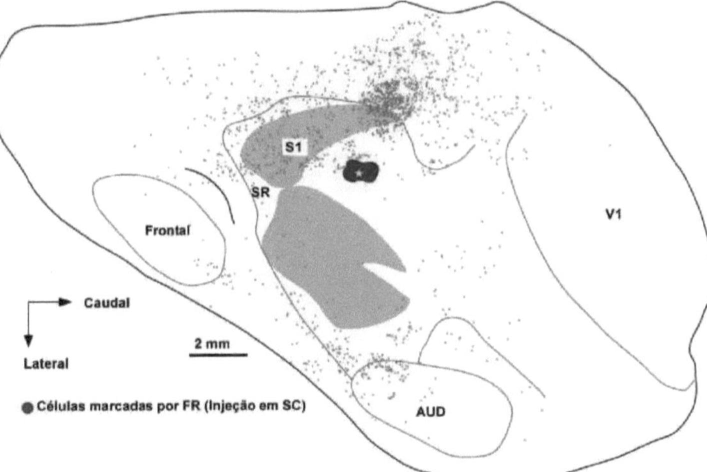

Figura 23.   Cortico-cortical projections for area SC (case 0705). Reconstruction of cross-sections of the neocortex. The FR injection was in the anterior paw representation of SC (a) and (b). In the anterior parietal cortex, most of the labelled cells were observed in the paw representation of Sl and in the region medial to Sl, SR and SC (c). Marked cells were also observed in S2/PV, auditory cortex, peristriate cortex, primary visual cortex, frontal cortex and medial zone of the neocortex.

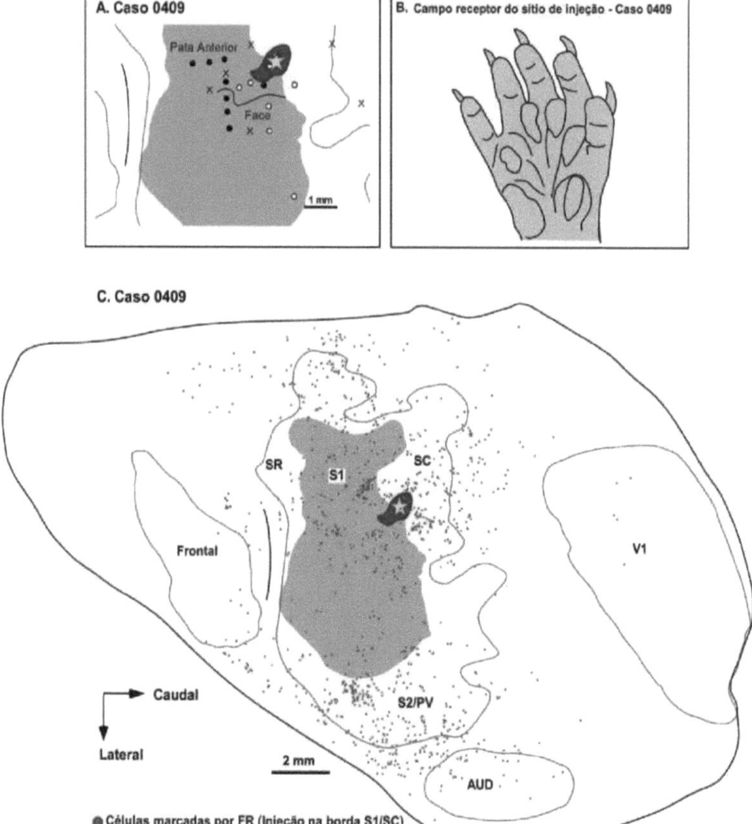

Figure 24. Cortico-cortical projections for area SC (case 0409). Cross-sectional reconstruction of the neocortex. The FR injection was positioned at the border between Sl and SC, in a region representing the anterior paw of SC (a) and (b). In the anterior parietal cortex, labelled cells were observed within the entire length of Sl, in SR and SC (c). Marked cells were also observed in S2/PV, auditory cortex, peristriate cortex, primary visual cortex, frontal cortex, insula and medial zone of the neocortex (c).

In case 0705, the auditory cortex was marked in its rostral half (Figure 23c), unlike case 0602, which had its caudal portion marked (Figure 22b).

Also in case 0705, few labelled neurons were observed in the primary visual cortex. A greater number of labelled cells were found in the peristriate cortex (Figure 23c).

The region medial to the parietal cortex again showed a high density of marked cells throughout its

70

anteroposterior extension, with those grouped above the SC and the peristriate cortex, which correspond to the paramarginal cortex, standing out (MARTINICH et al., 2000).

Additionally, labelled cells in the frontal cortex (anterior to the orbital fissure) were also found in case 0705 (but not in case 0602) (Figure 23c).

In ***experiment 0409***, an injection was made at the border between Sl and SC (Figure 24). The receptor field corresponding to the centre of the injection is illustrated in Figure 24b, and corresponds to the entire anterior paw (deep stimulation). The density of penetrations in this case was insufficient for the formation of a detailed somatotopic map of Sl, as well as for the electrophysiological delimitation of its rostral and caudal borders, which was estimated mainly by myelin marking.

The injection at the edge of Sl with SC differs from the other SC injections in that it shows a higher density of labelling in Sl (Figure 24c). In addition to Sl, the other labelled areas were: SR, SC, S2/PV, auditory cortex, peristriate cortex, VI, frontal cortex and medial region of the neocortex (Figure 24c). In these areas, the results obtained are similar to the injections carried out in the aforementioned cases.

4.1.4 Summary of the cortico-cortical projections for SI and SC.

The data described above show that area SI receives intrinsic cortical projections and projections from other somesthetic areas. These connections come from regions adjacent to the injection site in Sl, as well as areas SC and SR. Intense projections arising from the S2/PV area were identified in only one of the cases, where the injection site showed greater spreading.

In contrast, the caudal somesthetic area (SC) notably receives cortical afferents from different sensory modalities, since it receives projections from somesthetic, visual and auditory cortical regions, as well as projections originating from the frontal cortex. **42 Anatomical and functional characterisation of the sensorimotor thalamus**

This section will describe the thalamocortical projections for areas Sl and SC. The identification of thalamocortical projections helps to characterise the type of processing carried out in cortical areas. More specifically, the identification of projections from the motor and sensory thalamic nuclei to areas Sl and SC is important for clarifying the involvement of these areas in the sensory-motor circuit of the opossum *Didelphis aurita.*

42.1 Architectural subdivisions of the somesthetic and motor thalamus of the opossum

The identification of retrogradely labelled thalamic nuclei was carried out using coronal sections stained with cresyl-violet (Nissl method) and, in some cases, confirmed by sections reacted to reveal the enzyme cytochrome oxidase. The Nissl method relies mainly on cell body size, cell morphology and cell density to separate the different nuclei.

The architectural subdivisions of the somesthetic and motor thalamus (ventral thalamus) of the opossum *Didelphis aurita* previously described in the literature (OSWALDO-CRUZ E ROCHA-MIRANDA, 1968) were re-analysed in this study using the NADPH-diaphorase, cytochrome oxidase, parvalbumin and calbindin techniques. With the exception of Nissl, the aforementioned techniques had not yet been used to demarcate the thalamic nuclei of the opossum *Didelphis aurita.*

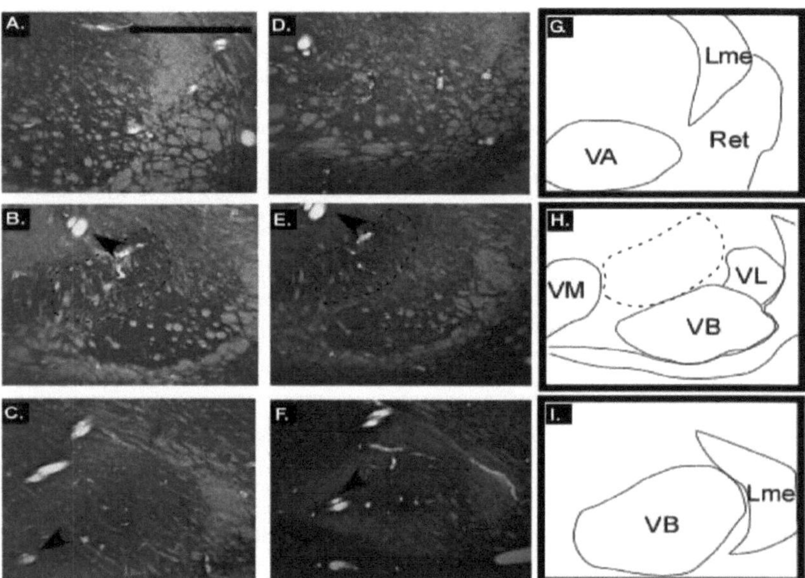

Figura 25.   **Marking pattern of cytochrome oxidase and NADPH-diaphorase enzyme activities in the thalamus.** The photomicrographs (A,D), (B,E), (C,F) are different planes in the rostro-caudal axis of the same animal. The left column illustrates the cytochrome oxidase labelling pattern, while the right column illustrates the NADPH-diaphorase pattern. The edges of the VA, VB and VL nuclei were easily identified by both methods and the same labelling pattern was observed in both markings, represented in the illustration (G, H, I). The sections illustrated side by side for the two different protocols are adjacent and the arrows indicate the

same blood vessel. The dotted lines in (B) and (E) correspond to a marking of the dorsal neuropil to the VB

consistently observed by both the cytochrome-oxidase and NADPH-diaphorase techniques and which is not

visualised with the other methods. Calibration bar = Imm.

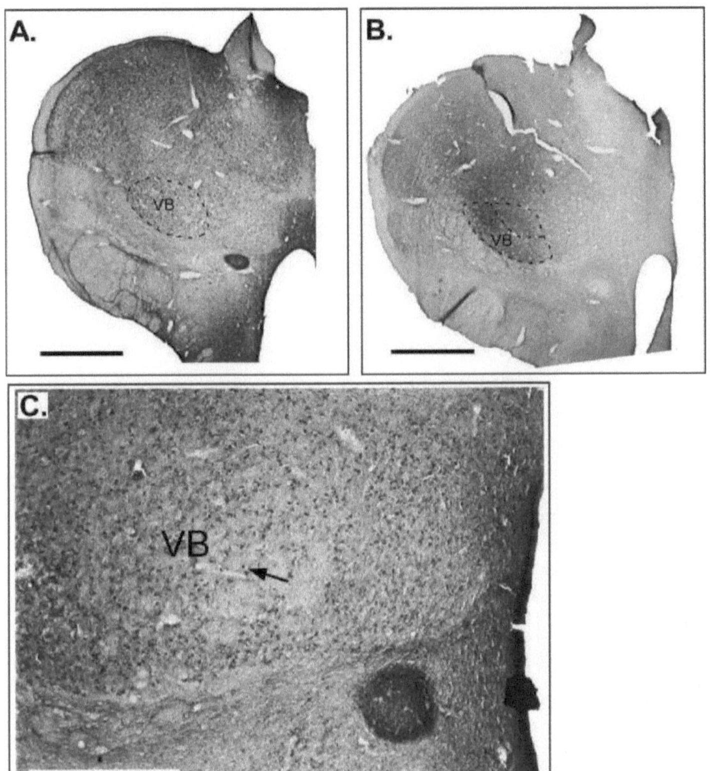

**Figura 26.   Thalamic subdivisions revealed by immunohistochemistry for calbindin and
parvalbumin. Note the edges of the VB thalamic nucleus in the section reacted for calbindin (A),**

and in (B) for parvalbumin. Calbindin-positive cells (arrow) were found diffusely

in the thalamus, including VB (C). Calibration bar = 500 μm (A and B); 125 μm (C).

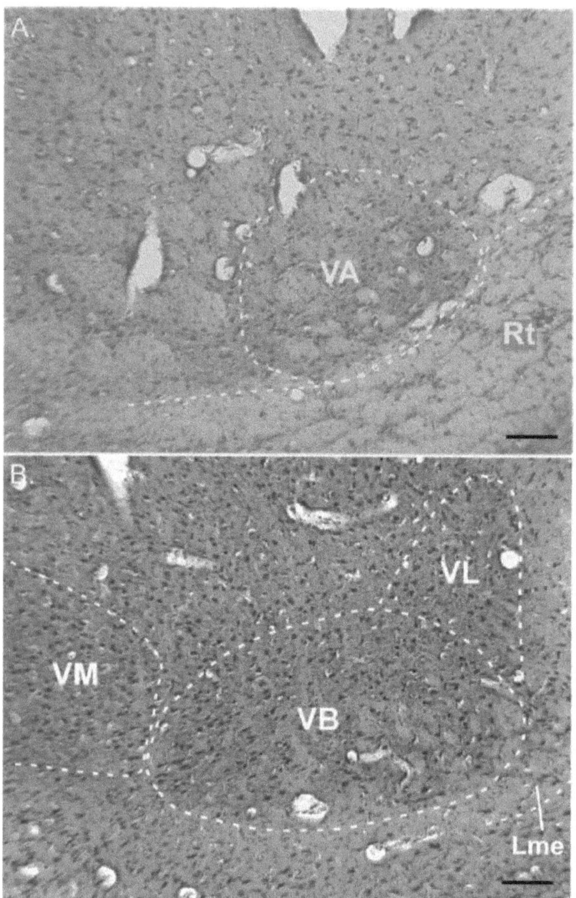

Figura 27. **Thalamic subdivisions revealed by the Nissl method.** In **(A)** **it can be** seen that the cells found in the VA nucleus are medium-sized, with a spiny and ovoid shape. In VB (B), the neurons are large or medium-sized with long cell processes, while in VL (B) the cells are also medium-sized, but polygonal in shape. Calibration bar = 200 μm.

The thalamic sections reacted for cytochrome oxidase and NADPH-diaphorase showed the same pattern of neurite labelling (Figure 25). The thalamic sections reacted for calbindin and parvalbumin differed in the pattern of their markings in the two cases analysed (case 0630 and 0631 - Figure 26). Notably, the thalamic sections reacted for calbindin showed many labelled calbindin-positive cells, which were distributed throughout the thalamus. In contrast, no parvalbumin-positive cells were visualised after reactions for parvalbumin in the two cases studied.

The dorsal thalamus in this species can be subdivided into several groups, such as: 1) anterior group, 2) midline, 3) medial, 4) ventral, 5) laterodorsal (OSWALDO-CRUZ and ROCHA-MIRANDA, 1968). The following description of the architectural subdivisions will be restricted to the ventral group, where the main somesthetic and motor thalamic nuclei are located.

4.12.1 Ventral basal thalamic nucleus (VB)

According to Oswaldo Cruz and Rocha-Miranda (1968), the nucleus (VB) of the opossum is limited to planes A = 5.5 and A = 3.2 and is characterised by being made up of large and medium-sized neurons with long cell processes. In our coronal sections, the VB nucleus was found rostrally at a level where the reticulated nucleus was no longer visible. At this level, the external medullary lamina (EML) can be seen, ventral to the VB (Figure 27). The dorsal limit of VB, however, is more difficult to delimit using the Nissl technique, since the nuclei neighbouring VB in this region have cells with similar morphology and density (Figure 27).

The sections reacted for CO and NADPH-diaphorase allowed the VB thalamic nucleus to be identified. This nucleus was intensely marked, characterised by intense reactivity for CO and NADPH-d (Figure 25). A nucleus or nuclear complex located dorsally to the VB was characterised by dense marking in the cytochrome-oxidase and NADPH-diaphorase reactions, but was not marked by either calbindin or parvalbumin.

The VB nucleus has the neuropil labelled for both calbindin and the parvalbumin reaction (Figure 26). The latero-ventral region of VB was more intensely labelled than the medial portion of the nucleus in both techniques. This marking suggests a subdivision of VB into two regions. These subdivisions were not clearly visualised with the other methods used (compare Figure 26 with 25 and 27).

In addition to the labelled neuropil described above, NADPH-diaphorase histochemistry revealed weakly labelled neurons in VB and in all the other labelled thalamic nuclei.

4.122 Ventral anterior (VA) and ventral lateral (VL) thalamic nuclei

The VL and VA nuclei form an xVL nuclear complex, which according to Oswaldo-Cruz and Rocha-Miranda (1968), extends from A= 6.7 to A= 4.4 and are identified in the most rostral cutting planes of the ventral group.

In the thalamus, VA appears as the first nucleus of the ventral group, located medially to the reticulate nucleus, and in the first plane of the section where the cerebral peduncle appears (Figure 27). In the Nissl preparation, its cells are medium-sized, spiny and ovoid in shape.

In the sections reacted for CO and NADPH-d, it was also possible to identify the "motor" thalamic nuclei (VA, VL and VM). VL was less reactive (Figure 25), allowing its ventral border to be identified with VA rostrally and with VB caudally. The neuropil of the VA nucleus was intensely reactive for both reactions (Figure 25).

The VL nucleus, at its rostral end, is located dorsally to VA, in a medial position in the first section where the external medullary lamina arises. This nucleus also has medium-sized cells, but with a polygonal shape (Figure 27).

4.1.23 Medial ventral thalamic nucleus (VM)

The VM nucleus forms a limited complex in the A= 6.3 and A= 4.0 planes (Oswaldo-Cruz and Rocha-Miranda, 1968). Its cell morphology repeats that described for VB, according to Nissl staining. This nucleus was identified as a group of cells located medially to the other ventral nuclei, located dorsally to the medial end of the zona incerta (Figure 27).

In the sections reacted for cytochrome oxidase and NADPH-diaphorase, the VM nucleus is well delimited and the neuropil is strongly marked (Figure 25).

In the sections reacted for calbindin and parvalbumin, the VM nucleus neuropil was weakly reactive or non-reactive (Figure 26).

Calbindin-positive cells are distributed throughout the dorsal thalamus. We observed a border separating the region of calbindin-positive cells from the region of non-marking cells. This border is at the boundary between the VB and the external medullary lamina (EML). Few calbindin-positive cells were visualised below this border (Figure 26c).

422 Thalamo-cortical projections for Sl and SC

422.1 Projection for the representation of the anterior paw in the primary somatosensory cortex (Sl)

A total of nine injections were made into the primary somatosensory cortex of five animals (Table 4).

In the rostral end of the primary somatosensory cortex, fluoro-emerald was injected in ***case 0602*** at 1.7

mm from the orbital fissure (Table 4) (Figure 20). Although this was a small injection (Table 4), it was possible to identify thalamocortical projections from the following groups: anterior group, ventral lateral group, laterodorsal group and medial group (Figure 28).

The few cells visualised in the anterior group were restricted to the ventral anterior (AV) and medial anterior (AM) nuclei (not shown). In the ventral lateral group, marked cells were found in the VA, VB, VL, VM and C nuclei, while in the laterodorsal group only the ventral intermediate lateral nucleus (Liv) was marked. In particular, the cells found in VB were scattered throughout the nucleus without obvious somatotopic segregation. In some sections, the labelled cells appeared to be concentrated in the latero-ventral region of the nucleus (Figure 28 - A 4.90 and 4.50). The nuclei in the medial group that were marked were the paracentral (PC) and mid-dorsal (MD) nuclei.

Quantification in this case (Figure 30) revealed that the VB nucleus had a higher incidence of labelled cells (59.5%) than the VA (1.5%) and VL (42%) motor nuclei.

In contrast to this pattern, the injection of DY at the border with the SC area (Figure 29) marked fewer cells in the thalamus than the injection of FE. The nuclei of the anterior group were not marked by the DY injection. The few marked cells were restricted to the dorsal region of the VA, VL, Liv and C nuclei.

Considering that this was the second largest injection carried out in our study (Table 4), it is impossible to relate this restricted pattern of thalamic projections to Sl with the volume of the injection. The next experiment to be described suggests that this result is a consequence of the slow transport of diamidino-yellow. Therefore, it is possible that the animal's survival time was insufficient for effective retrograde labelling of cells in the thalamus. Another possibility is that the cortical lesion caused by the insertion of the crystals damaged the thalamic afferents, preventing their transport to the thalamus.

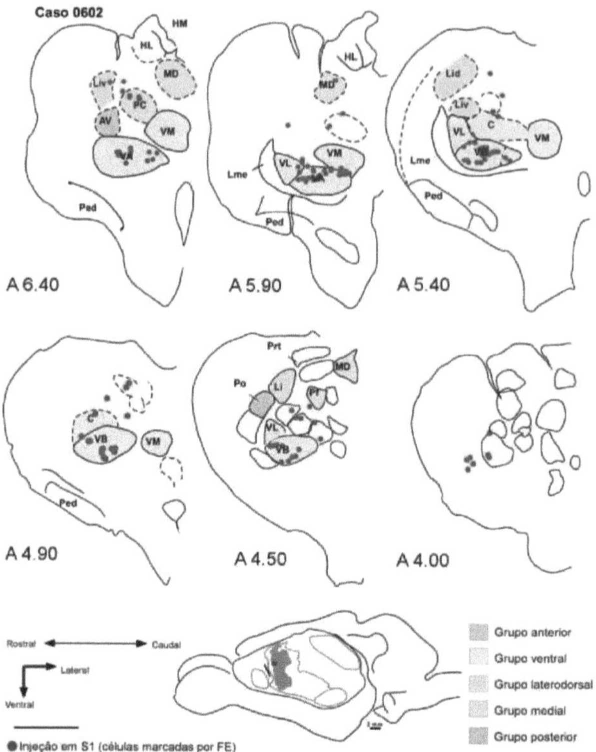

Figure 28. Thalamo-cortical projections for area Sl (case 0602). Reconstruction of coronal sections of the thalamus. Above, schematic drawings of sections of the thalamus with the nuclei and the distribution of cells marked by the neurotracer FE, injected at the border of areas SR and Sl (below). The thin black lines denote the borders between the thalamic nuclei. The dashed lines represent less obvious borders of the thalamic nuclei marked with Nissl staining. The numbering next to the thalamic sections corresponds to the section plan of the opossum stereotaxic atlas (OSWALDO-CRUZ and ROCHA-MIRANDA, 1968). In this case, the FE injection was placed in the representation of Sl's anterior paw (Figure 20a and b). Most of the labelled cells were observed in the VB and VA nuclei. Few labelled cells were visualised in the VL, VM, PC and C nuclei.

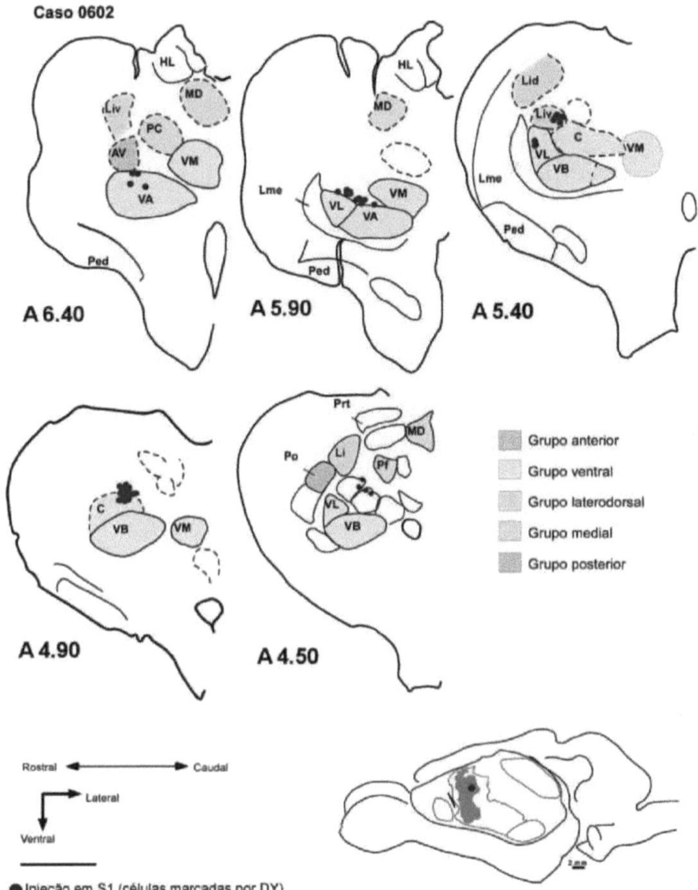

Figura 29. **Thalamo-cortical projections for the SI area (case 0602).** Reconstruction of coronal sections of the thalamus.In this case, the DY injection was placed in the anterior paw representation of SI (Figure 10a and b). Few labelled cells were visualised in the VA, VL, Liv and C nuclei.

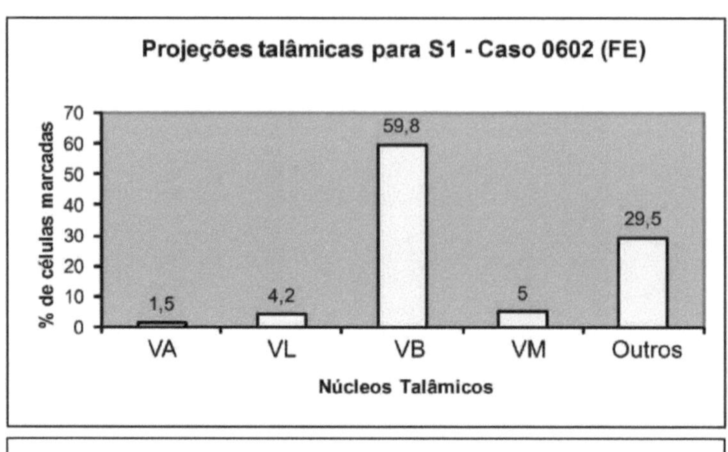

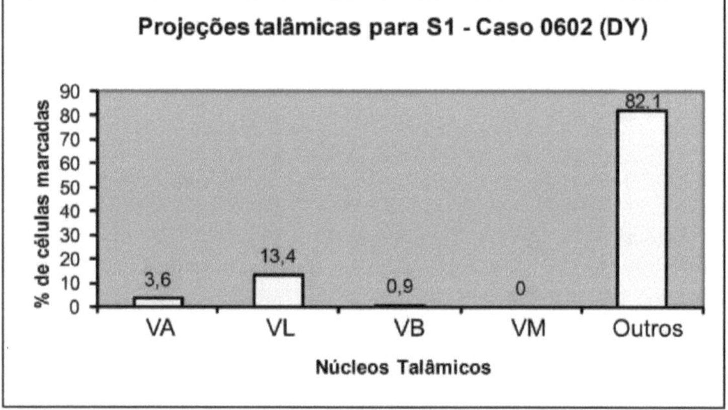

Figura 30. Quantitative analysis of thalamic projections for somatotopic representation

paw in Sl (case 0602) - FE and DY injections. Note that the VB somesthetic nucleus had a higher density of labelled cells in the FE injection, followed by the VL nucleus.

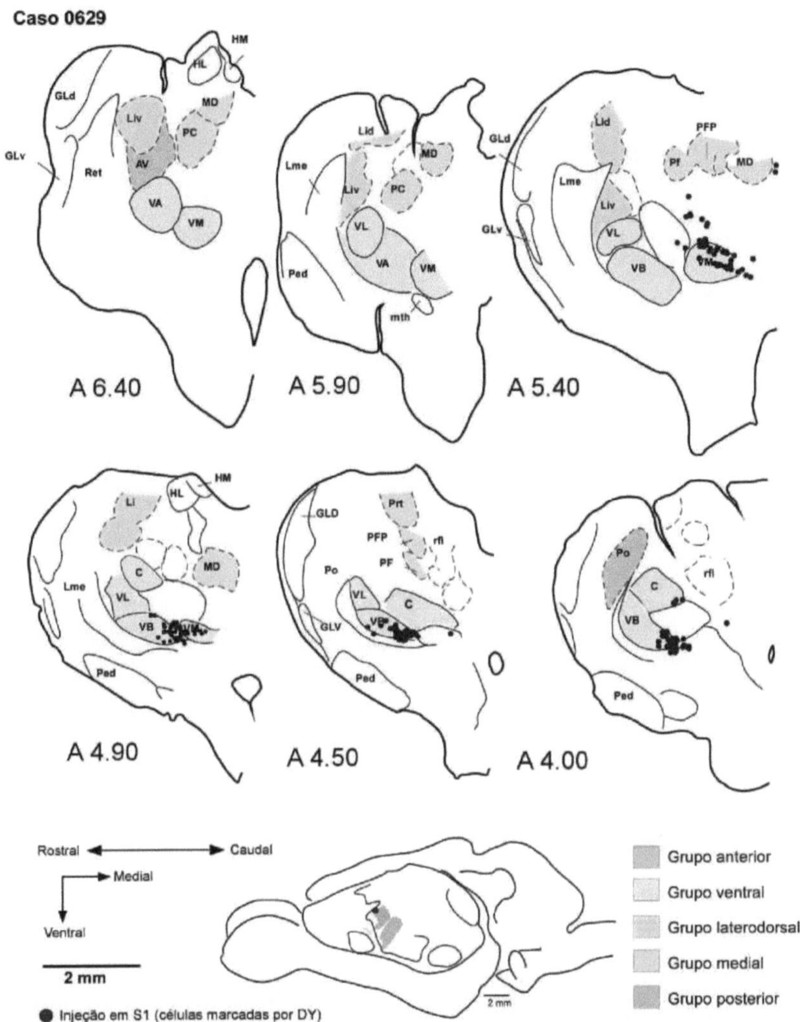

Figura 31. Thalamo-cortical projections for area Sl (case 0629). Reconstruction of coronal sections of the thalamus. In this case, the DY injection was placed in the representation of the anterior paw of Sl. Most of the labelled cells were observed in the VB, VA and PC nuclei. Few labelled cells were visualised in the VB, VM and C nuclei.

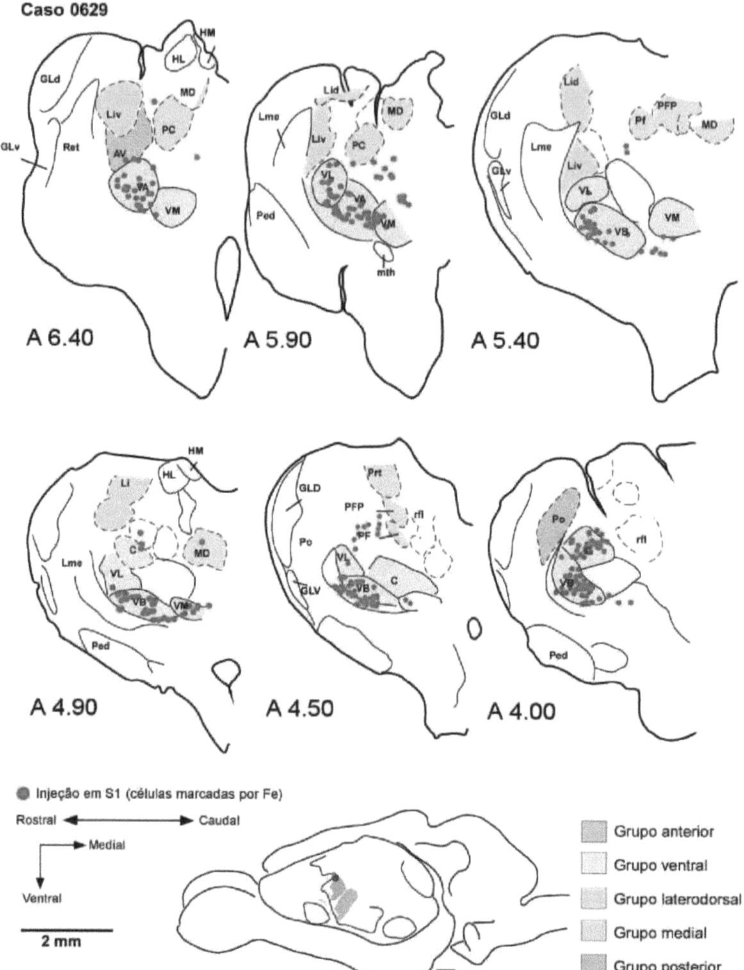

Figura 32. Thalamo-cortical projections for area Sl (case 0629). Reconstruction of coronal sections of the thalamus. In this case, the FE injection was placed in the representation of the anterior paw of Sl. Most of the labelled cells were observed in the VB, VA and C nuclei. Few labelled cells were seen in the AV, VL, VM, PF, PFP, PC and MD nuclei.

In case 0629, two injections were made in Sl with the neurotracers DY and FE (Table 4 and Figure 33). The DY injection was positioned 2.5 mm from the orbital fissure and its injection site covered around 0.9677 mm^2 of the primary somatosensory cortex.

Again, after the DY injection, labelling was restricted to a few nuclei in the thalamus. These results can be seen in Figure 31, which shows the absence of cells in nuclei from the previous group (Figure 31, A

82

7.00 - A 5.90). In fact, the neurons marked by DY were mainly limited to the medial region of the VB nucleus (A 5.40-A 4.00, 65.3% of the neurons marked, Figure 31); and to a lesser extent, to the neighbouring VM eC nuclei.

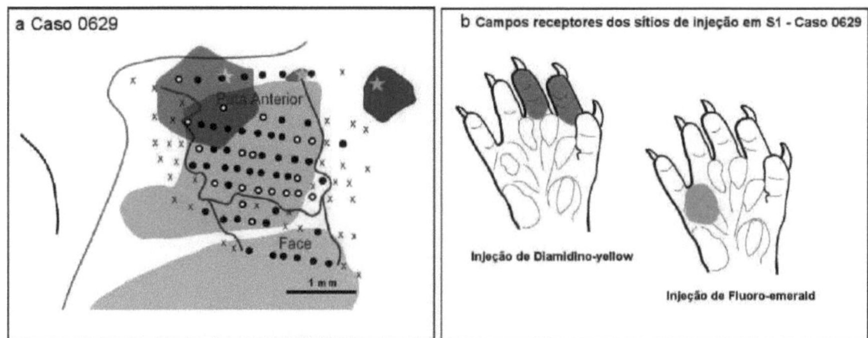

Figura 33.   Injection sites of case 0629. (A) shows the DY, FE and FR injection sites (blue, green and red circles respectively). The receptor fields for the penetration sites closest to the injection sites are small (B).

In this same case, we performed an injection of FE in Sl at 3.6 mm from the orbital fissure (Table 4). Analysis of this injection revealed several marked thalamic nuclei (Figure 32), such as: AV in the anterior group; VA, VL, VB, VM and C in the ventral lateral group; parafascicular (PF), posterolateral parafascicular (PFP), PC and MD in the medial group and the posterior nucleus (Po) in the posterior group. Once again, the VB nucleus was more intensely marked (34.5%), followed by VA (17.2%) (Figures 32 and 34).

In experiment 0705, projections of the thalamic nuclei were also studied for the representation of the anterior paw in Sl. The DY site showed an area of 1.3545 mm^2 , expanding into areas SR and SC (Figure 21A). In the thalamus, the following nuclei were marked (Figure 35): AV, AM (anterior group), VA, VB, VL, VM, C (ventral lateral group), Liv (laterodorsal group) and PF, PC (medial group). A region in the medial zone of the thalamus was marked, but the nuclei could not be identified by the architectural methods used.

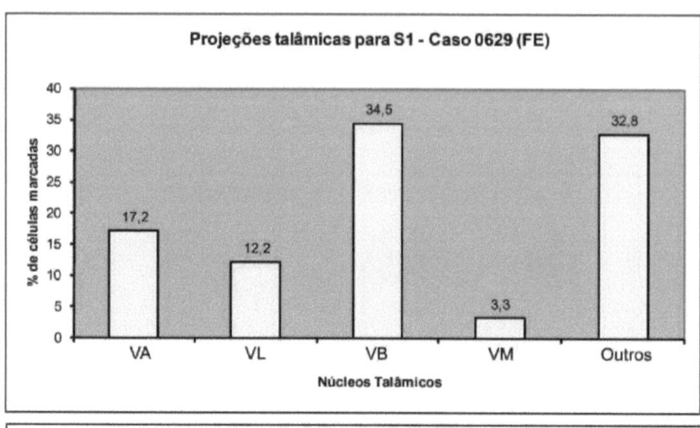

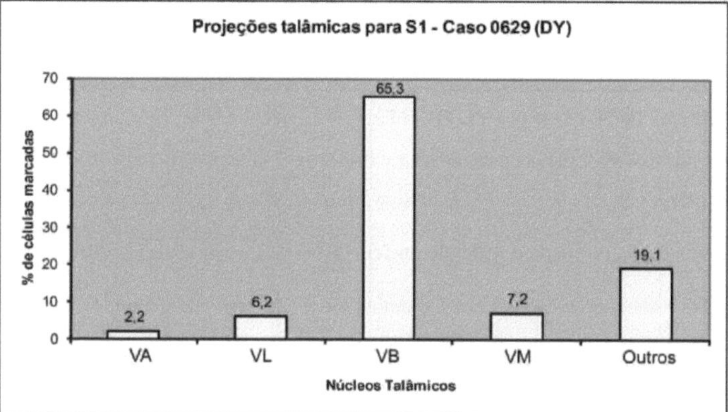

Figura 34.   Quantitative analysis of the thalamic projections for the somatotopic representation of the paw in Sl (case 0629) - FE and DY injections. Note that the VB somesthetic nucleus had a higher density of labelled cells than the motor nuclei in both FE and DY injections.

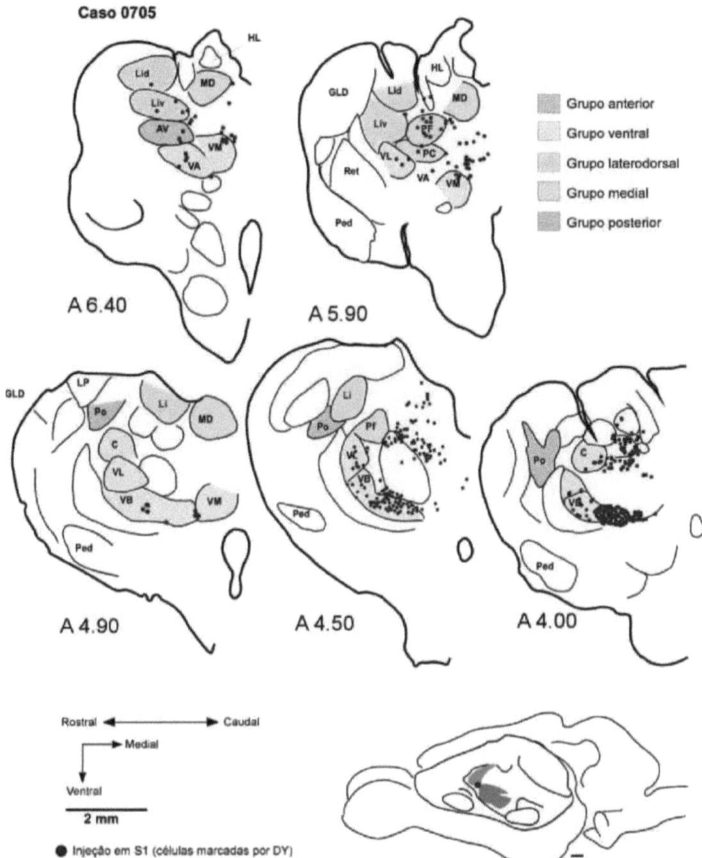

Figura 35. Thalamo-cortical projections for area SI (case 0705). Reconstruction of coronal sections of the thalamus. In this case, the DY injection was placed in the representation of the anterior paw of Sl, with spreading of the neurotracer to areas SR and SC. The labelled cells were observed in the AM, AV, VB, VA, VL, VM, C, PF and PC nuclei.

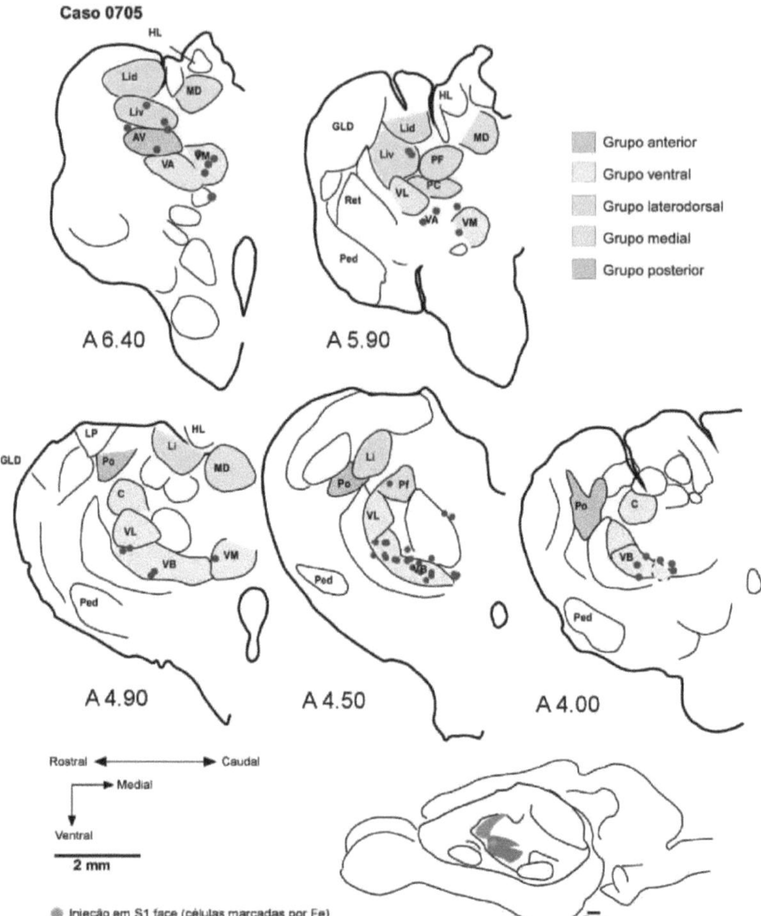

Figura 36. **Thalamo-cortical projections for area Sl (case 0705).** Reconstruction of coronal sections of the thalamus. In this case, the FE injection was placed in the representation of the face of Sl. Most of the labelled cells were observed in the VB, AM nuclei. The remaining labelled cells were visualised in the AD, VA, VL and VM, Liv and PF nuclei.

42.22 Projection for the representation of the face in the primary somatosensory cortex

In experiment 0705, an injection of FE was performed in the SI face representation (Figure 36). The distribution pattern of the labelled thalamic neurons was similar to that obtained after injections in the anterior paw representation (Figure 35). Retrogradely labelled neurons were identified in the following groups: anterior group (AV, AM, AD nuclei), ventral lateral group (VB, VA, VL and VM nuclei), dorsal lateral group (Liv) and medial group (PF). Although the overlap of the injected site with the result of our mapping suggests the

86

absence of tracer spread for the representation of the anterior paw (Figure 20A), we found labelled cells along the entire length of VB (Figure 36, sections A4.90 and 4.50); although in the more caudal sections, labelling in the VB nucleus was found more medially (Figure 36, A 4.00) where the representation of the face has been electrophysiologically identified in this species (SOUZA et al., 1971).

In case 0705, a DY injection was also made in the representation of the front foot. This injection showed a spread to the region representing the SI face, which can be seen in Figure 21. As illustrated in Figure 35, we noticed marked cells in the lateral region of the nucleus, but the retrogradely marked cells were concentrated in the medial region of the VB nucleus in the posterior sections (A 4.50 and 4.00), reinforcing the existence of a topographical organisation within VB.

423 Summary of thalamic afferents for SI:

It was found that the following thalamic groups project to the representation fields of the foreleg and SI face: 1) the anterior group, 2) the ventral group, 3) the laterodorsal group and 4) the medial group.

The consistent projections to SI, which appear in all cases in the anterior group, originate in the ventral anterior nucleus (AV). In the ventral lateral group, the projections arise from the ventral anterior motor nucleus (VA), ventral medial nucleus (VM) and ventral lateral nucleus (VL). The somesthetic component of the ventral group originates from the ventral basal somesthetic nucleus (VB) and the central nucleus (C). Neurons from the medial group with projections to the anterior paw and face representations were identified in the parafascicular nuclei (PF), paracentral nucleus (PC), while projections from the laterodorsal group arise from the ventral intermediate lateral nucleus (Liv).

Marginally topographic projections from the thalamus to SI were found in the VB nucleus. The cells that project to the face representation of the primary somesthetic cortex tend to be concentrated in the medial region of the posterior pole of the nucleus; even so, labelled cells were found throughout the entire length of the nucleus, regardless of the region injected.

1.1 .2.4 Thalamo-cortical projections for CS

Four injections were made into the SC in four different animals (Table 4), making it possible to characterise the pattern of thalamo-cortical projections to this area. In all cases, the injection of FR into the SC resulted in numerous marked cells in the thalamus, including, in one of the cases analysed (0705, see below), the marking of thalamic nuclei that had not been marked by the injections made in Sl.

Figure 37 shows case 0602, where the injection site is located 5.1 mm from the orbital fissure (Table

4). This was the most caudal injection performed in our experiments.

The resulting thalamic labelling (Figure 37) was found in the AV, AM (anterior group), VA, VB, VL, VM, C (ventral lateral group), Liv, dorsal intermediate lateral (Lid) (laterodorsal group) PF, PC, MD (medial group) and suprageniculate (SG) nuclei (posterior group).

Among them, the nucleus with the highest concentration of labelled cells was the VB nucleus (22.4%) (Figure 40). A lower incidence of cell labelling was also noted in motor nuclei VL (12.2%) and VA (4.2%).

In contrast to case 0602, the injection of FR into the SC in case 0629 (Figure 38) revealed a slightly more restricted labelling, found in the following nuclei: AV (anterior group), VA, VB, VL, C (ventral lateral group) and Po (posterior group). It should be noted that the area occupied by this injection site (Table 4) is larger than in case 0602, which would not explain this limited labelling in the thalamus.

Quantifying the projections to the SC again revealed a predominance of projections originating from the VB nucleus (39.8 per cent) over the VA and VL motor nuclei (16.6 and 16 per cent, respectively) (Figure 40).

In ***experiment 0705,*** SC injection revealed neurons in previously unlabelled thalamic nuclei, such as: AD in the anterior group and Li in the laterodorsal group (Figure 39). The other nuclei that showed projections to the SC in this case were: AV, AM (anterior group), VA, VB, VL, VM, C (ventral lateral group), Lid, Liv, Li (laterodorsal group), PC, PF, MD (medial group) and Po (posterior group).

1.4   4 Summary of thalamic afferents to SC: Projections to area SC originate from the following thalamic groups: 1) the anterior group, 2) the ventral lateral and laterodorsal groups, 3) the medial group and 4) the posterior group. The nuclei of the anterior group that send consistent projections to SC include the medial anterior nucleus (AM) and the ventral anterior nucleus (AV). In the ventral lateral group, the motor nuclei that project to the SC correspond to the ventral anterior nucleus (VA), ventral medial nucleus (VM) and ventral lateral nucleus (VL). The ventral basal somesthetic nucleus (VB) also projects to the SC. The central nucleus (C) also projects to the SC. In the laterodorsal group, the nuclei that project to the SC are: ventral intermediate lateral (Liv) and dorsal ventral intralaminar (Lid). Neurons from the medial group with projections to the SC were identified in the parafascicular (PF), paracentral (PC) and mid-dorsal (MD) nuclei. Finally, the nuclei of the posterior group are highlighted as they were not frequently identified in the SI injections. These are the posterior nucleus (Po) and the suprageniculate nucleus (SG).

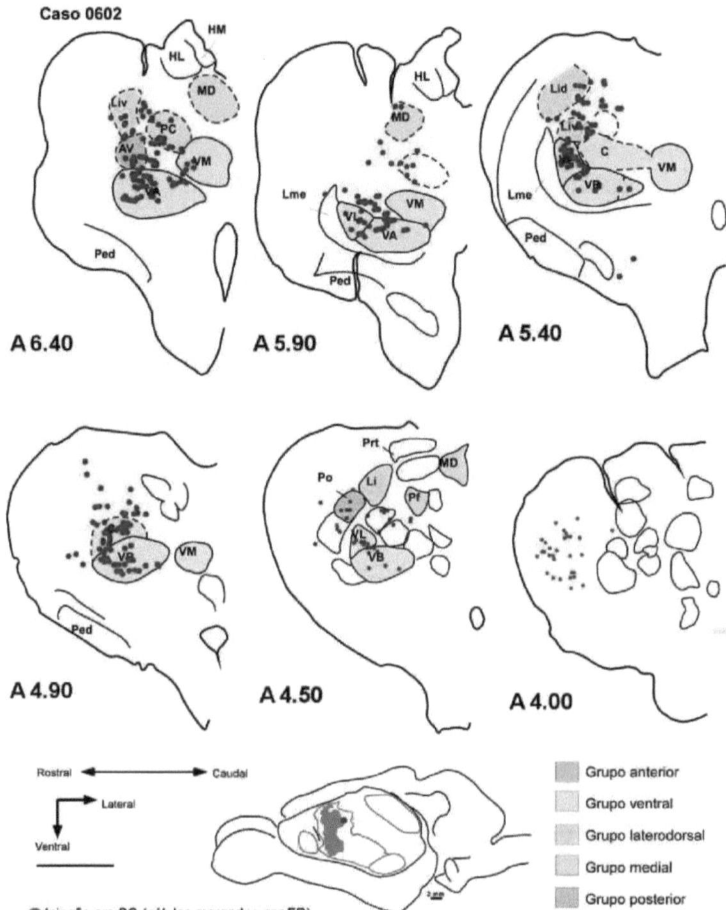

Figura 37. **Thalamo-cortical projections for area SC (case 0602).** Reconstruction of coronal sections of the thalamus. The FR injection was positioned in SC. Most of the labelled cells were seen in the VB, VA and AV nuclei. The remaining labelled cells were visualised in the AV, AM, VA, VB, VL, Liv, Lid, C, PF, PC, SG and MD nuclei.

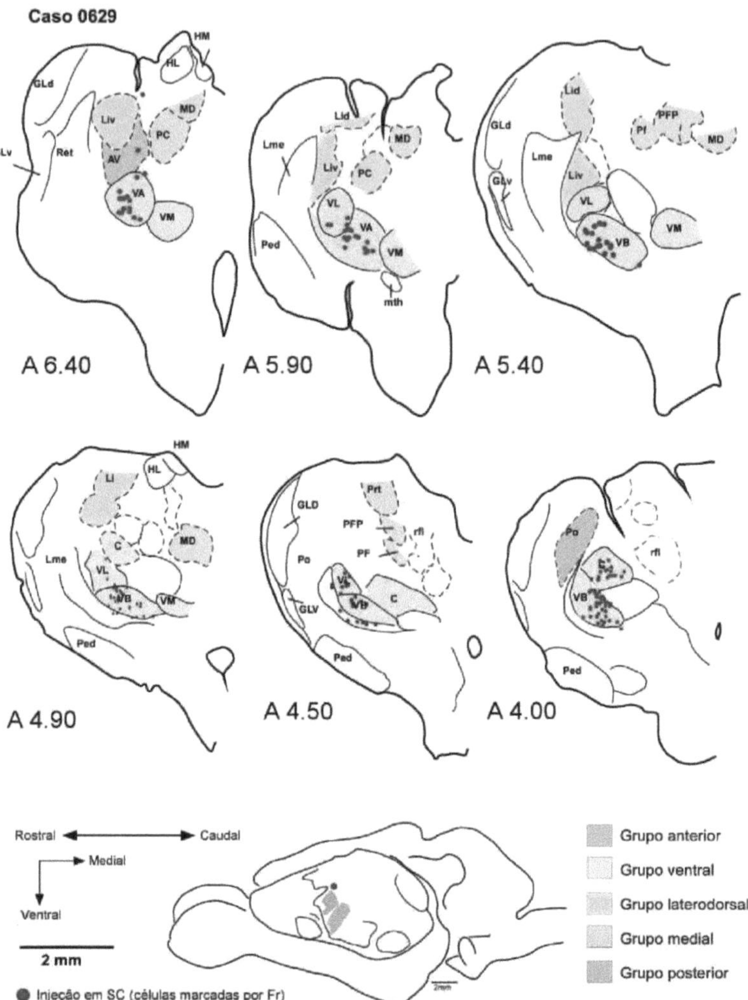

Figura 38. Thalamo-cortical projections for area SC (case 0629). Reconstruction of coronal sections of the thalamus. In this case, the FR injection was positioned in the SC. The labelled cells were visualised in the AV, VA, VB, VL, C and Po nuclei.

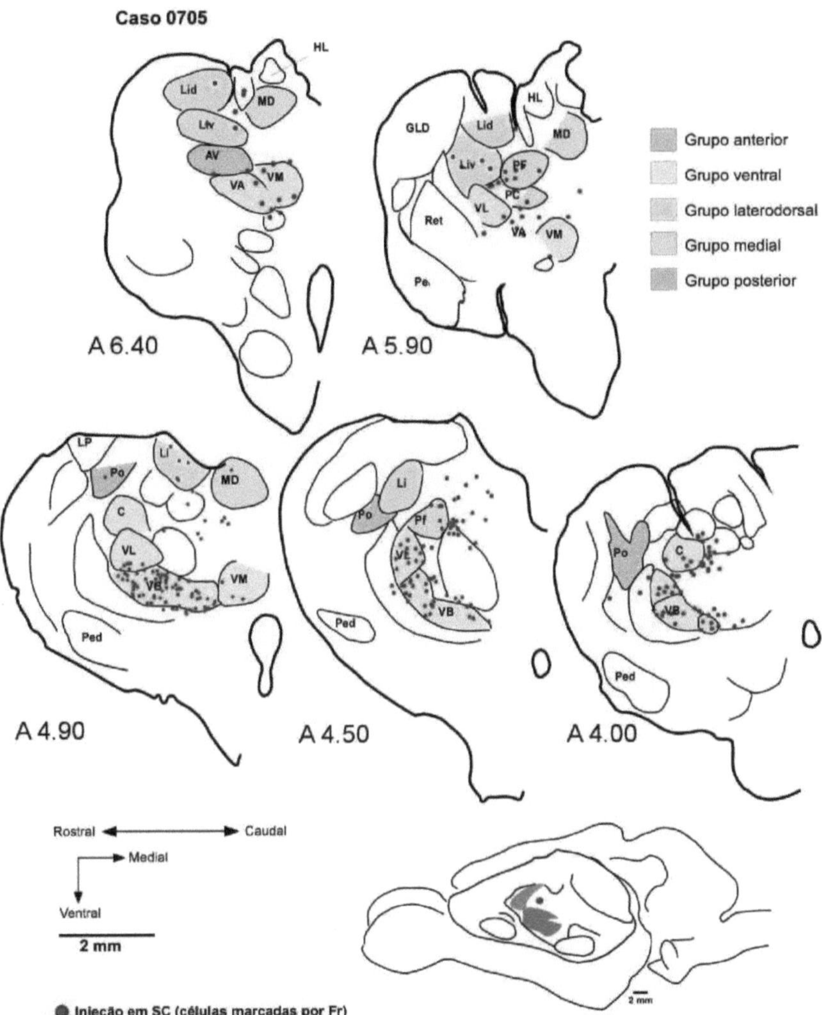

Figura 39. Thalamo-cortical projections for area SC (case 0705). Reconstruction of coronal sections of the thalamus. The FR injection was positioned in SC. Most of the labelled cells were seen in the VB, VL and AM nuclei. The remaining labelled cells were visualised in the AV, AM, VA, VB, VL, VM, Lid, Liv, AD and Li nuclei.

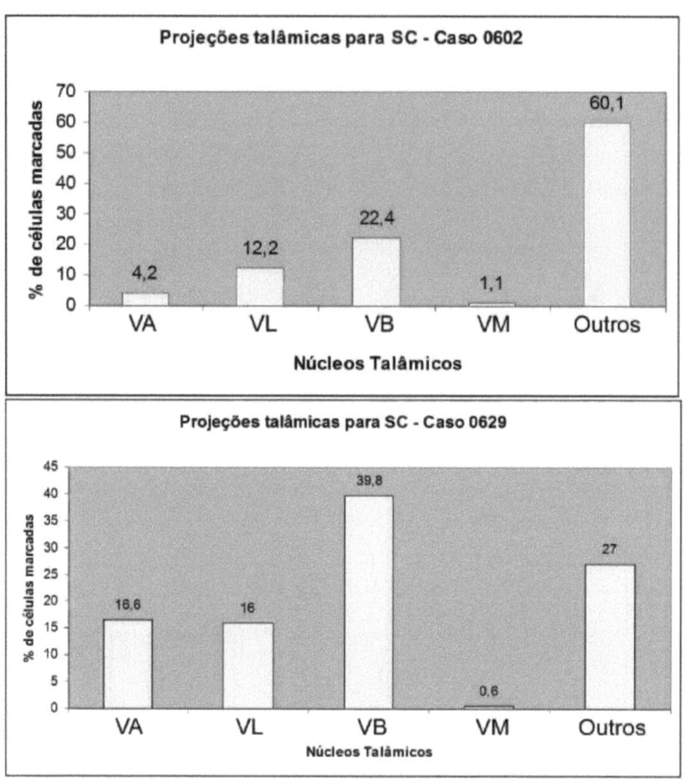

Figura 40. Quantitative analysis of the thalamic projections for the SC representation. It can be seen that the VB somesthetic nucleus had a higher density of labelled cells in both cases. The VM nucleus was poorly marked in these cases analysed for SC injections.

CHAPTER 5

DISCUSSION

5.1 The neocortex and opossum behaviour

There is currently a trend in comparative neurobiology to characterise the somatosensory areas caudal to the S1 in primates, including those that make up the posterior parietal cortex (PADBERG et al., 2005; PADBERG and KRUBITZER, 2006, STEPNIEWSKA et al., 2005). This effort is partly due to the proposed interrelationship between the evolution of manual skills and the organisation of somatosensory and motor areas in the neocortex.

In general, low-complexity movements are a common feature among didelphids (KIMBLE, 1997). Although opossums are fast and efficient when catching prey, they perform very stereotyped movements when handling prey. Other non-primate mammals, such as rats, are able to perform more complex movements during predatory behaviour, using their limbs to manipulate food by pronating, supinating and adjusting their fingers to reach the food (IVANCO et. al., 1996).

Although comparative studies indicate that mammals share some common characteristics in the organisation of the somatosensory and motor system, the motor skills of the hands and other limbs seem to be related to anatomical differences, such as those in the endings of the pyramidal tracts in the spinal cord of these animals. The anatomical differences in the endings of the cortico-spinal tract make it possible to classify mammals into four distinct groups (IVANCO et al., 1996). In the first group, where the opossum is found, the cortical fibres end in the cervical and thoracic spinal cord, mainly in the dorsal part of the contralateral grey matter and the intermediate zone (IVANCO et al., 1996). In the second group, where we observed the rats, the fibres are distributed throughout the spinal cord and end contralaterally in both the dorsal and intermediate zones of the spinal cord. A few cortical fibres from group II are also distributed to lamina IX, which contains motoneurons in the ventral part of the spinal cord *(cf.* MARTIN and FISHER, 1968). In the other two groups, which include some primates such as humans, the cortico-spinal fibres end densely in lamina IX of the ventral como *(cf. MARTIN* and FISHER, 1968). Thus, in the opossum group and other marsupials, the endings of the cortico-spinal tract in the medial part of the dorsal como of the spinal cord resemble sensory rather than motor projections (IVANCO et. al., 1996, BECK et al., 1996).

It is now known that the corticospinal tract has more than one site of origin in at least 22 mammalian species. In most species, these sites appear predominantly on the side contralateral to the tract (90%). In marsupials, the main source of these projections corresponds to the parietal cortex or sensorimotor amalgam (NUDO and MASTERTON, 1990). In primates, the main projection zone includes Brodman's areas 6, 4, 1, 2, 5 and part of areas 23 and 24. In this way, it can be seen that the somatosensory areas also share consistent projections to the pyramidal tract, so that electrical stimulation of sites in these areas is capable of producing body movements even after complete Ml damage (NUDO and MASTERTON, 1990). If we take into account the distribution of these projections, it has been observed that in marsupials, the distribution of these projections in the parietal cortex tends to be homogeneous, while in primates the region rostral to the somatosensory cortex, corresponding to the motor cortex, has denser projections. This denser projection region is located parallel to the central sulcus, within area 4, and also includes the border with area 3a (NUDO and MASTERTON, 1990). The existence of corticospinal projections from the somesthetic cortex is not, therefore, a unique parameter for characterising this area as having motor functions in the opossum.

It is believed that other characteristics may be involved in the behaviours and movements of different species, such as the peripheral morphology of the limbs. The organisation of the motor cortex is also directly related to the motor ability of a species. Sections of the rat's pyramidal tract or lesions of the motor cortex result in impaired reaching movements with the limb contralateral to the lesion. Injured animals use a series of rotational body movements to reach for food without moving their hands, but they do not abduct their elbows or adduct their paws to bring the food to their mouths. These movements resemble the reaching strategy of opossums, which do not perform elbow abduction or paw pronation or supination.

The presence or absence of a primary motor cortex that controls motor functions in the opossum *Didelphis aurita* will be discussed below. The main focus of this study was to try to understand how the somatosensory areas relate to the execution of movement in this species through their thalamic and motor connections. We will first discuss the segregation of motor and somatosensory information at subcortical levels, such as the thalamus, and then analyse the cortical connections themselves.

5.2 Delimitation of the thalamic nuclei

The somesthetic and motor thalamic nuclei of the opossum *Didelphis aurita* were previously described using the classic Nissl and Myelin reactions by Bodian (1939), Oswaldo-Cruz and Rocha-Miranda (1979) and Donoghue and Ebner (1981). Our architectural assessment of some thalamic nuclei aimed to re-analyse and confirm the boundaries between the somesthetic and motor nuclei previously proposed for *Didelphis aurita*. The chosen techniques of NADPH-diaphorase, calbindin and parvalbumin were recently used successfully in the opossum *Didelphis aurita* to re-analyse the nuclear borders of the amygdaloid complex (ROCHA-REGO et al., 2008).

5.2.1 Histochemical markers

Both the NADPH-diaphorase and cytochrome-oxidase reactions proved to be good markers of the thalamic nuclei of the opossum *Didelphis aurita,* and the borders between the VB (somesthetic), VL, VA and VM (motor) nuclei of the ventrai group were identified with both methods. In addition, a correspondence in the pattern of neurite labelling between NADPH-diaphorase and cytochrome-oxidase was observed. This colocalisation of NADPH-diaphorase marking with cytochrome oxidase reinforces the link between the activity of these two enzymes (see also FRANCA, 1999). The cytochrome oxidase technique reveals the activity of a mitochondrial enzyme, cytochrome aa3. Regions intensely marked by this histochemical method therefore correspond to portions of the nervous tissue that are metabolically more active (WONG- RILEY, 1979), where greater neuronal activity is recorded *(cf.* LIVINGSTONE and HUBEL, 1984; ALLMAN and ZUCKER, 1990).

5.2.2 Immunohistochemical markers

In the opossum *Didelphia aurita,* the distribution of calbindin- and parvalbumin-positive cells in the thalamus differs. Notably, numerous calbindin-labelled cells are scattered throughout the dorsal thalamus, while parvalbumin-positive cells appear to be non-existent. In both cases 0630 and 0631 we observed a reactive neuropil for calbindin and parvalbumin. However, these markers did not stand out in the delineation of the somesthetic and motor thalamic nuclei.

Historically, two types of thalamocortical connections have been proposed with different roles in the cortical activities of primates (JONES, 1998). The first group of connections would arise from the main thalamic transmission nuclei (including the sensory ones), with a topographically organised projection to the middle layers of the cerebral cortex. The other group originates in the intralaminar nuclei, and possibly the associative nuclei. Their projection is diffuse and directed towards the superficial layers of the cerebral cortex, indicating involvement in generalised aspects of brain function. It is believed that these neurons that project to the superficial layers of the cortex are distributed throughout the dorsal thalamus without respecting nuclear borders. The neurons projecting to the middle layers of the cortex are organised in a more restricted manner, mainly evident in the sensory and motor transmission nuclei.

Evidence for these two systems is based on studies of the distribution and connection of cells in the thalamus of monkeys by immunohistochemistry for calbindin and parvalbumin-binding proteins (JONES, 1998). In these studies, it was observed that calbindin-positive cells extend throughout the nuclei, while parvalbumin-positive cells are restricted to a few specific nuclei.

It is possible that the pattern of cellular marking shown in this study indicates the existence of diffuse thalamic projections to the cortex of the opossum *Didelphis aurita* (JONES, 1998). However, relating the labelling of calcium-binding proteins in isolation to the existence of thalamic projection systems in certain species could lead to errors, since the specific expression of these proteins and the connection of their cells with the cortex has not been systematically studied in non-primate mammals such as the opossum (JONES, 1998).

5.2 Electrophysiological mapping of the anterior parietal cortex of the opossum *Didelphis aurita*

Assuming that the somatosensory and motor systems of *Didelphis aurita* are segregated in the thalamus into somesthetic and motor nuclei, we need to analyse the organisation of these two systems in the neocortex. Based on the proposal that the primary motor cortex overlaps with the primary somesthetic cortex in this species (LENDE 1963), we will begin by analysing the electrophysiological characteristics and borders of Sl, and then discuss the cortical and thalamic projections to areas Sl and SC.

The somatosensory cortical surface in the anterior parietal cortex of *Didelphis aurita,* as proposed by Beck et. al (1996), has subdivisions that present different patterns of connections. Our results indicate that the somatosensory areas Sl, and SC in the opossum *Didelphis aurita* differ by cortico-cortical and thalamo-cortical projections, although in the former these differences are clearer. The presence of these areas has been observed in other mammals, such as the opossums *Didelphis virginiana and Monodelphis domestica,* the *rat*

(Euarchontoglires rat), the tupaia *Tupaiinae* (REMPLE et al., 2006) and the porcupine (KAAS, 2004).

In general, the somatotopic representation of S1 studied reflects the pattern described for other mammals, with the anterior paw represented medially, followed by the face laterally (BECK et al., 1996, MAGALHÃES-CASTRO and SARAIVA, 1971, HUFFMAN and KRUBITZER, 2001). Our study was the first to show a detailed representation of the S1 toes of the opossum *Didelphis aurita.* In other marsupials, detailed analysis of the internal organisation of S1 shows a crude representation of the forepaw in S1. In *Didelphis azarae azarae,* cortical regions representing the forelimb and hand were found, but not the fingers individually (MAGALHÃES-CASTRO and SARAIVA, 1971). In *Didelphis virginiana,* the somatotopic representation of S1 on the medial-lateral axis and the reversal of receptor fields between S1 and S2 were focussed on (BECK et al., 1996). Little detail was revealed about the individual representation of the fingers. Our results also suggest that the facial vibrissae are individually represented within S1 (Figure 15).

On the other hand, in the opossum *Didelphis aurita* it was not possible to adequately map the SR and SC areas. In the two anaesthetic protocols tested throughout this study, both areas were predominantly non-responsive to somatosensory stimulation, although sites responsive to somatosensory stimuli could be observed in both SR and SC. Sites responsive to somatosensory stimulation were also observed in the SR and SC areas of *Didelphis virginiana* (BECK et al., 1996), providing electrophysiological evidence that these areas process somatosensory information. Previous electrophysiological studies in *Didelphis azarae azarae* did not consider the existence of these areas (MAGALHÃES-CASTRO and SARAIVA, 1971) and, therefore, we cannot confirm their existence in didelphids other than those mentioned.

The differences in myelination observed between areas S1, SR and SC have also been described in *Didelphis virginiana* (BECK et al., 1996). In primates, such as the marmoset, area 3b, homologous to S1 in non-primate mammals, appears highly myelinated between area 3a rostrally (slightly marked) and area 1 caudally (moderately marked) (KRUBITZER and KAAS, 1990). In bats, area 3b is coextensive with zones of dense myelination (KRUBITZER et al, 1993). Again, the region of areas 1/2 was slightly marked, as was area 3a (KRUBITZER et al, 1993). In the *flying fox,* the posterior parietal cortex was shown to be a weakly myelinated region between V2 (moderately myelinated) and areas 1/2.

5.3 Cortical projections for Sl and SC

Based on our results, it can be seen that in the opossum *Didelphis aurita the* somesthetic areas are connected to each other. We observed that area Sl receives exclusively somesthetic projections, originating in the rostral (SR), caudal (SC), secondary somesthetic/parietal ventral (S2/PV) somesthetic areas (Figure 40). Intrinsic projections of adjacent representations and those coming from the SR and SC areas predominate in all the cases analysed. On the other hand, contrary to expectations, projections from the S2/PV area were very sparse or even absent in most of the cases analysed. It is possible that the small diameter and volume of our injections was insufficient to show a greater density of these projections. The presence of numerous retrogradely labelled cells in S2/PV after the insertion of a DY crystal in the representation of the anterior paw in Sl, in case 0705, seems to confirm this possibility (Figure 23). In the opossum *Didelphis virginiana*, connections between areas Sl, SR and SC have been previously demonstrated (BECK et al., 1996). Neurotracers injected into SI showed irrelavant labelling in S2/PV.

In primates, such as marmosets, these connections include areas of the motor system (KRUBITZER and KAAS, 1990). In this species, area 3b reciprocally connects with areas 3a, 1, S2 and PV. Less dense projections are also observed with areas 2, Ml, the supplementary motor area (SMA) and the limbic cortex medial to SMA (KRUBITZER and KAAS, 1990). In rats, SI connections include areas of the parietal and frontal cortex, such as adjacent regions of Sl, S2, PV, PM, PR, PL, Agl and Agm - the latter two corresponding to the motor cortex and the supplementary/pre-motor area, respectively (FABRI and BURTON, 1991).

On the other hand, area SC exhibited a pattern of cortico-cortical projections that included a greater number of cortical areas. Although this area appears to be part of the somatosensory system in the opossum, our results provide strong evidence of **multimodal processing** in this region, with connections originating in the somesthetic cortex (Sl, SR, SC, S2/PV), visual cortex, auditory cortex and frontal cortex (Figure 40).

Often, the areas immediately caudal to area 3b have been hastily described as homologous to somatosensory area 1, although this idea is based mainly on their anatomical position, and not on electrophysiological criteria and their projections. However, caudal to Sl, or area 3b in different species, one can find several areas that can only be identified as a somatosensory area, posterior parietal cortex or visual area by their projections and/or electrophysiological characteristics.

In non-primate mammals, such as squirrels, insectivores and some marsupials, the caudal cortex to Sl has neurons that respond both to deep stimulation of contralateral receptors and in some cases to visual stimulation *(cf.* SLUTSKY et al., 2000; KRUBITZER et al., 1997; HUFFMAN et al., 1999; FROST et al., 2000). This region is known as the caudal field (C), caudal somatosensory area (SC), medial parietal area (PM) or area *Vi (cf.* PADBERG et al., 2005).

In rats, the medial parietal area (MP) occupies the same position as the SC (caudal to the SL) and, as in the opossum *Didelphis aurita,* also receives projections from the visual, auditory and somatosensory cortex (REMPLE et al., 2006).

The tupaia (Tree shrew) is a squirrel-like mammal and a close relative of primates. This species also has area SC caudally to 3b, which differs architecturally from 3b by a less dense granular layer (REMPLE et al., 2006). The cortico-cortical projections of this area include Ml, M2, the medial motor area, somatosensory areas (3a, 3b, S2, PV) and subdivisions of the posterior parietal cortex (REMPLE et al., 2006).

Caudal to area SC, the tupaia has a posterior parietal cortex. Injections into the posterior parietal cortex of this species retrogradely marked V2, VI, S2, PV, SC, subdivisions of the posterior parietal cortex and areas of the temporal cortex (REMPLE et al., 2006). Areas 3b, 3a and Ml had very few markings (REMPLE et al., 2006). These cortico-cortical projections of the tupaia's posterior parietal cortex are similar to our injections in area SC in that they include markings in the visual, somatosensory and temporal areas.

In primates, the posterior parietal cortex has expanded considerably. In galagos, the rostral half of the posterior parietal cortex, caudal to the SC, receives dense projections from S2 and PV, and projects to the premotor cortex (KAAS, 2004). This possible specialisation of the rostral region of the posterior parietal cortex of the galago clearly differs from the multimodal posterior parietal cortex of the tupaia (Tree shrew).

In anthropoid primates, especially Old World primates, the posterior parietal cortex has several subdivisions (KAAS, 2004). The more rostral fields receive somatosensory projections from area 2 and the lateral parietal cortex and project to Ml and other motor areas. The more caudal areas receive auditory and visual information and project to premotor areas (KAAS, 2004).

When we analysed the somatosensory areas located caudally to area 3b in primates and rostrally to the posterior parietal cortex, we observed little similarity with our injections into area SC in the opossum *Didelphis aurita.* In the primate *Callicebus moloch,* area 1 connects consistently with areas 3b, 2, S2/PV, 5, 7b/AIP, and sparsely with areas 3a, M1 and the frontal cortex. No connection with any other sensory area has been observed in this species (PADBERG et al., 2005).

In the macaque monkey, another somatosensory area caudal to area 1 has been identified: area 2. Neurons in area 2 are very responsive to cutaneous and deep receptor stimulation in both awake and anaesthetised animals (PADBERG et al., 2005). This area has a complete representation of the contralateral receptor surface, just like area 3b and 1, but with larger receptor fields. Injections made in area 2 indicate connections with areas 3b, 1, 3a, S2, M1 and area 5 (PADBERG et al., 2005). As with area 1, no cortical connections are observed between area 2 and other sensory modalities.

The comparison of cortical connections between these species and our results supports the hypothesis that the existence of the posterior parietal cortex, as observed in large primates (KAAS, 2004) may have originated from a single multimodal area. The projections of area SC of the opossum *Didelphis aurita* more closely resemble the posterior parietal cortex of primates and other mammals than other areas immediately caudal to Sl, such as areas 1 and 2.

5. 5 Thalamo-cortical projections for Sl

In the opossum *Didelphis aurita,* retrograde marking in the thalamus was similar between the cases analysed, although some variations were present. Small variations in the thalamic projections to the somesthetic cortex have also been observed in primates, such as the zogue *(Calicebus moloch)* (PADBERG and KRUBITZER, 2006). As our results illustrate, the primary somesthetic area of the opossum *Didelphis aurita* receives consistent thalamic projections from the VB nucleus. However, motor thalamic nuclei also project to Sl according to this study. These nuclei are the ventral anterior (VA), ventral lateral (VL) and ventral medial (VM) thalamic nuclei.

Thalamic projections to Sl have been extensively studied in primates. Undoubtedly, the main source of projections to area 3b is the VP nucleus, both in New World and Old World primates *(cf.* JONES et al., 1979). In New World primates, the VP nucleus is the main source of thalamic projections to 3b, although less dense projections also originate in the anterior pulvinar nuclei, and in some cases in VPS, VL and CL (KRUBITZER and KAAS, 1992).

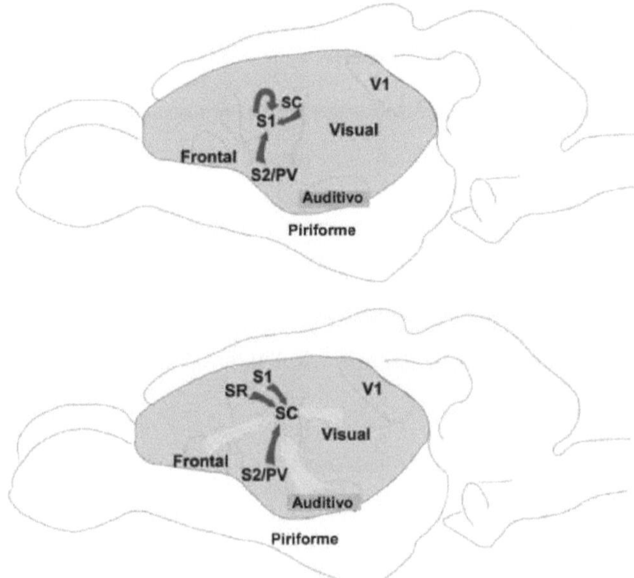

Figure 41. Schematic drawing of cortical afferents to areas Sl and SC. Area Sl receives projections from other cortical areas, such as SR, SC, S2 and PV. Area SC receives projections from the somesthetic cortex, other sensory areas (visual and auditory) and the frontal cortex. The red arrows illustrate the somesthetic connections, while the yellow arrows represent the connections from other sensory modalities or the frontal cortex.

In galagos, these projections are more widespread and include VP, VIM, VL, CM, CL and PO (PEARSON and HAINES, 1980). However, these results stem from very large injection sites that may also have included the motor cortex (PADBERG and KRUBITZER, 2006). Projections of nuclei homologous to VP to Sl have been widely demonstrated in other mammalian groups such as rodents, carnivores and other primates (KAAS, 1983). In *Didelphis aurita, it can* be seen that the cells that project to the anterior paw representation of Sl spread throughout the VB nucleus, and not exclusively from its lateral region. Few signs of topographical segregation were observed in these thalamocortical projections. It is possible that the medial marking in VB after injection in the representation of the face reflects the somatotopic representation of the nucleus previously described by electrophysiological recording (SOUZA et al., 1971). In this representation, the rostro-caudal axis of the animal's body coincides with the medio-lateral axis of the nucleus respectively. In other mammals, the ventro-posterior complex has a subdivision representing the face medially called the medial ventroposterior nucleus (VPM), while the rest of the body is represented more laterally, in the lateral ventroposterior nucleus (VPL) (KAAS, 1988). In our results, only these subdivisions seem to be evident only in the Parvalbumin reactions.

The projections of the motor thalamic nuclei to Sl were previously described by Donoghue and Ebner (1981) in *Didelphis virginiana[1]* , although in this study, the Sl area defined by the authors also included SR and SC. According to Donoghue and Ebner (1981), when neutracer injections were made in Sl, the VB, VL, CIN and CL nuclei were marked.

However, there are important differences in the architectural criteria used to delimit the VL nucleus between our study and Ebner's (DONOGHUE and EBNER, 1981). The atlas of the opossum brain by Oswaldo-Cruz and Rocha-Miranda (1968), used as a reference in our work, was the only one to illustrate the three "motor" nuclei identified in other mammals (VL, VA and VM) (BODIAN, 1939; OSWALDO-CRUZ and ROCHA-MIRANDA, 1979). The nucleus to which this study refers (DONOGHUE and EBNER, 1981) as VL apparently consists of the VA nucleus of Oswaldo-Cruz and Rocha-Miranda (1968) and the architectural description previously made by Bodian (1939). According to these two studies, the VA nucleus coexists medially with the reticulated nucleus and is later replaced by the VB nucleus. This location corresponds exactly to what Donoghue and Ebner (1981) identify as VL. In our study, we adopted the criteria of Bodian (1939) and Oswaldo-cruz and Rocha-Miranda (1979) for identifying the VL nucleus.

These differences in the delimitation of these thalamic nuclei seem to explain the differences between our results and those of Donoghue and Ebner (1981). The latter considered a larger portion of the thalamus to correspond to the VB nucleus, and therefore observed lateral retrograde labelling in the nucleus after injection into the parietal cortex (DONOGHUE and EBNER, 1981). The VB nucleus defined in this study occupies a smaller region of the thalamus and corresponds only to the lateral portion of VB mentioned by Donoghue and Ebner (1981). For this reason, after the injection of retrograde neurotracer into the anterior paw of Sl, our study found neurone labelling throughout the VB nucleus, while Donoghue and Ebner found it only in the lateral portion of the nucleus, inferring that this labelling represents a topographical segregation of the anterior paw.

Analysing afferent projections to these thalamic nuclei is an enlightening alternative that can be used to understand their subdivisions. Somesthetic projections from the gracilis and cuneiform nuclei via the medial lemniscus to the thalamus (WALSH and EBNER, 1973) are profusely directed towards the lateral part of VB mentioned by Donoghue and Ebner. Somesthetic projections originating in the trigeminal nucleus (corresponding to the face) end in the VB nucleus, predominantly in the posterior pole of the nucleus (WALSH and EBNER, 1973). The lateral portion of VB that receives somesthetic projections corresponds exactly to the VB nucleus in Oswaldo-Cruz and Rocha-Miranda's atlas (1968). In our study, this region was retrogradely marked after injection into the cortical representation of the Sl. anterior paw.

On the other hand, axons originating in the deep nuclei of the cerebellum travel to the dorsal region of VB (WALSH and EBNER, 1973). This region corresponds to the VL nucleus of the atlas of Oswaldo Cruz and Rocha-Miranda (1968). Earlier, projections of the same cerebellar nuclei are found ending in the region that corresponds to the VA nucleus of the atlas by Oswaldo Cruz and Rocha-Miranda (1968), strengthening the interpretation of the thalamic subdivisions proposed by these authors.

In cases 0602 and 0629, the VL projections to the anterior paw representation of the primary somesthetic cortex tend to originate topographically from the ventral region of the nucleus, close to its border with VB, although retrogradely labelled cells were also found in the dorsal portion of the nucleus in some sections. In other mammals, such as cats and monkeys, the ventral region of VL projects into the posterior zone of the cortex. This projection zone is related to distal musculature, while the dorsal part of VL projects to anterior zones representing proximal musculature.

In addition, nuclei from other thalamic groups project to the primary somesthetic cortex in the opossum *Didelphis aurita*. In the anterior group, the main projection nucleus to this area seems to be the ventral anterior nucleus (AV), although projections have been found from the dorsal anterior nucleus (AD) and medial anterior nucleus (AM). According to Bodian (1939), in the North American opossum *(Didelphis virginiand)*, lesions in the retrosplenial and cingulate cortex were able to induce degeneration in the lateral portion of the dorsal anterior (DA) and ventral anterior (VA) nuclei. Individual projections from AV seem to be directed towards the insula and cingulate cortex. The AD nucleus, on the other hand, would send projections to the retrosplenial and post-orbital areas. Finally, the AM nucleus would be responsible for connections with an interhemispheric region of the frontal pole.

5.6 Thalamo-cortical projections for SC

In our study, both the VB somesthetic nucleus and the VA, VL and VM motor nuclei project to the SC area. In addition to these nuclei, we observed consistent projections from the AV, AM, Liv, PF, PC and MD nuclei to SC. There appears to be a convergence of projections from some thalamic nuclei to areas Sl and SC in the opossum *Didelphis aurita*. The major differences between these projections lie in the greater spread of labelled cells after injections in SC.

Although the SC area of the opossum *Didelphis aurita* receives projections from the same somesthetic and motor thalamic nuclei as Sl, there is a difference in the proportion of VB projections to both areas. More than half of the projections to Sl come from the VB nucleus (with the exception of cases where DY injections were made). Similarly, in the SC area, projections originating in VB also predominate, but are proportionally smaller (Figure 34 and 40). Of the three Sl injections analysed, in two of them more than 50% of the marked cells were in the VB nucleus, while in the SC injections we observed only 22.4 or 39.8% of the marked cells in VB.

Other thalamic nuclei that stand out in projections to SC because they don't project to Sl are the posterior group nucleus Po and the medial group nucleus MD.

In the opossum *Didelphis virginiana,* the region called posterior parietal cortex by Donoghue and Ebner (1981) and which receives projections from VL, VM, CIN and CL, is located caudally to the VB projection zone. In the opossum *Didelphis aurita,* the region of the cortex immediately caudal to the Sl receives multimodal cortical afferents and could therefore be homologous to the posterior parietal cortex of primates. Another possibility is that there is a region located between SC and the visual cortex in the opossum *Didelphis aurita that is homologous to* the posterior parietal cortex of primates. However, such a subdivision, if it exists, could not be evidenced in our study.

5.7 Sensorimotor amalgam hypothesis

The concept of the sensory-motor amalgam consists of the primary somatosensory cortex (SL) overlapping with the primary motor cortex (Ml) in a single area of the parietal cortex - the somato-motor cortex

- and was proposed for the opossum *Didelphis virginiana* by Lende (1963). In addition, the main characteristic of the sensory-motor amalgam consisted of the fact that the same cortical region responsible for sensory processing of a certain region of the body corresponded topographically to the cortical region responsible for motor control of that same part of the body. According to this proposal, the presence of a sensory-motor amalgam in a mammal with primitive cortical characteristics suggests that the first ancestral mammals had an equally amalgamated somatosensory and motor cortex (Figure 42a). Thus, the spatially separated SI and MI areas that we currently identify in eutherian mammals could have arisen from the duplication of cortical tissue corresponding to the sensory-motor amalgam of ancestral mammals (LENDE 1963, 1964). In order to discuss this hypothesis, it is necessary to evaluate what evidence would confirm the existence of the MI area in this animal, and if so, where it would be located.

The results found in our study, **if considered in isolation,** corroborate the hypothesis of the sensorimotor amalgam in the opossum. Throughout this discussion we will refer to the region of the sensorimotor amalgam as "motor SI" for methodological reasons: the absence of sufficient evidence to confirm the existence of a primary motor cortex. The use of this term does not conceptually imply that there is no sensory-motor amalgam in this region, as proposed by Richard Lende in 1963. We are just being cautious not to affirm the presence of a motor area in the parietal cortex as proposed by Lende, since there is not enough evidence to confirm this.

As we have seen, the consistent projections from the VA/VL motor nuclei to SI support the hypothesis of information processing from motor centres in this region (BECK et al., 1996) (Figure 42 a and b). However, the main source of thalamic projection to SI originates from the VB nucleus, a thalamic nucleus that receives tactile information from the sensory periphery via the gracilis and cuneiform nuclei (KAAS, 1983, 2004). And as we saw earlier, this region has the electrophysiological and histological characteristics of a primary somesthetic area, such as those identified in eutherian mammals (Figure 42b). In primates, area 3b has densely grouped cells in layer IV and III and VI (KAAS, 1983).

Evolution of the primary motor cortex in mammals

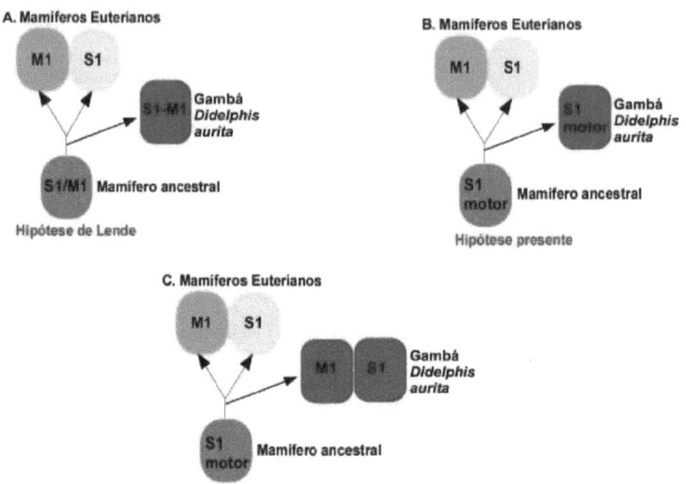

Figura 42. Schematic drawings of hypothesised evolution of the Sl and Ml cortices in mammals. According to Richard Lende (1963), the first ancestors had a sensorimotor amalgam, which would have been retained by opossums (A). Over the course of evolution, the eutherian mammals would have had separate areas Sl and Ml from the amalgam. However, electrophysiological records subsequent to Lende's do not confirm the amalgam hypothesis. The architectural organisation of the amalgam region shows an identifiable layer IV, suggesting that the amalgam is a primary somesthetic area (BECK et al., 1996, Gray, (B). It is possible that the primary motor area has not yet been identified in the cortex rostral to Sl (C). However, this region in the opossum seems to receive somesthetic thalamic projections and has the architectural organisation characteristic of a sensory area (Modified from BECK et al., 1996).

In the opossum, the parietal cortex has a granular layer IV and in no way resembles the agranular motor cortex (BECK et al., 1996; GRAY, 1924). In addition, electrophysiological mapping of area 3b in primates shows that the responsive sites in this area respond to superficial cutaneous stimulation, as observed in the primary somesthetic cortex of the opossum *Didelphis aurita,* where most of the sites are responsive to superficial cutaneous stimulation (KAAS, 1983).

Furthermore, the electrophysiological data indicating the existence of Ml in the parietal cortex of the opossum remains controversial. The first mapping studies used less refined electrophysiological techniques to identify Ml, using electrostimulation with superficial electrodes (LENDE, 1963). Later studies using microelectrostimulation, where the electrode is thinner and penetrates the cortex, did not confirm the existence of Ml completely amalgamated with Sl in the opossum. Beck et al. (1996) did not find a primary motor area topographically coincident with Sl in *Didelphis virginiana* after microelectrostimulation of the parietal cortex with low currents. Another study in *Monodelphis domestica* showed only a partial overlap of the motor cortex with the somesthetic cortex in face representation (FROST et al., 2000).

Considering another position for Ml in the opossum's neocortex seems a difficult task, as Lende himself eliminated the possibility of a motor region characteristic of Ml in the frontal lobe (LENDE, 1963, 1964) (Figure 42c). In his work, stimulation of this region only resulted in eye movements. The eye movements, variations in pupil diameter, retraction of the nictating membrane and eye closure that he describes after electrical stimulation of the frontal cortex are characteristic of the so-called ocular frontal field of placental mammals (LENDE, 1963).

The cortico-spinal projections of the opossum *Didelphis virginiana* are characterised by a few projections that originate from the region rostral to the region of the amalgam (NUDO and MASTERTON, 1990) (Figure 42c). Most of the projections that make up the cortico-spinal tract of this species originate in S1, and even so, these projections are directed towards the posterior aspect of the medulla.

Analysing Sl's cortical architecture also casts doubt on the presence of a motor cortex in the parietal cortex, as identified in other mammals. As we have seen, this region has a granular layer IV characteristic of sensory areas (GRAY, 1924; BECK et al. 1996). However, SR also has a well-developed granular layer IV, suggesting a sensory function (GRAY, 1924; BECK et al., 1996).

Martinich (1996) investigated the thalamocortical projections to the region located around the orbital fissure. Donoghue and Ebner (1981) also carried out injections in the region immediately posterior to the orbital fissure in the opossum and observed labelled cells in the medial nuclei PF and PC, the ventral nuclei VB and VL and the intralaminar nuclei. In *Didelphis aurita,* we injected the SR area and observed retrograde labelling in the VB, VA, VL and VM thalamic nuclei (unpublished data). As in Sl and SC, there was a predominance of cells originating in VB projecting to SR, weakening the possibility that this region could correspond to the primary motor area in the opossum (ANOMAL, 2003).

If we turn our attention to other mammals looking for cortical organisations that resemble the sensorimotor amalgam, we will see that other marsupials seem to have such an organisation, such as the *brush-tailed opossum* (REES and HORE, 1970), the central plateau *opossum (Didelphis azarae,* MAGALHÃES-

CASTRO and SARAIVA, 1971) and *Monodelphis domestica* (FROST et al., 2000). In the case of the cuíca, it is suggested that only the representations of the mandible and vibrissae form a sensory-motor amalgam. According to this proposal, the cortico-spinal projections originate in area SI of all these species and not in the frontal cortex (NUDO AND MASTERTON, 1990).

Neuroanatomical techniques (analysing projections) in other marsupials also indicate the existence of a sensory-motor amalgam, albeit with a partial overlap. According to axonal degeneration studies, the motor cortex of the *quokka wallaby, western grey kangaroo* and *red kangaroo* is located caudal to the post-orbital area, where we find the sensorimotor amalgam in the opossum (WATSON, 1971).

However, the Ml area has been spatially separated from SI in non-marsupial mammals. In less specialised placentals, there is still a partial overlap of the somatotopic representation of SI and Ml, but in most placental mammals these representations are completely separate and mirror each other (KAAS, 2004, 2008; LARSEN and KRUBITZER, 2008).

In monotremes, various theories have been proposed, but apparently there is one representation spatially coincident with SI and another located earlier in a mirrored position (LARSEN and KRUBITZER, 2008).

In birds, the circuit that promotes voluntary movements seems to involve thalamic nuclei comparable to the ventral thalamus of mammals (MEDINA et al., 1997), although the precise homology between them has not yet been completely established. Indicating similarities with the thalamocortical connections of SI and SC in the opossum *Didelphis aurita,* in the rostral Wulst of the penguin, there are thalamic projections originating in the anterior ventral intermediate dorsal somesthetic nucleus - DIVA (MEDINA et al., 1997), which is the main target of the projections of the dorsal column nuclei *(cf.* KORZENIEWSKA and GÚNTÚRKÚN, 1990). The rostral Wulst of birds has been compared to the Sl and Ml areas of mammals, and electrophysiological studies have confirmed that this region makes up a somatosensory region *{cf* FUNKE, 1989). Strengthening the idea that the rostral Wulst may be comparable to the Ml, projections are observed in penguins from the dorsointermediate (DIP) and ventrointermediate (VIA) thalamic nuclei. The VIA nucleus, in turn, receives afferents from the basal nuclei, indicating the existence of a pallido-thalamo-cortex pathway mediating motor functions (MEDINA et al., 1997). The VIA nucleus receives additional projections from the substantia nigra and the internal nuclei of the cerebellum and remembers the motor part of the thalamic nuclei, such as VA, VL and VPL (MEDINA et al., 1997).

But are the VIA and DIVA nuclei of the dorsal thalamus of birds homologous to the VA, VL and VP nuclei of mammals? If these nuclei are homologous, one would expect comparable regions in the thalamus of reptilian ancestors.

In turtles, a somatosensory field appears to be present in the dorsal part of the cortex. In lizards, the somatosensory field is located in the rostral pallium, which receives projections from the dorsal thalamic nucleus comparable to the DIVA of birds and VP of mammals. This region of the dorsal thalamus receives projections from the spinal cord and dorsal column nuclei (HOOGLAN, 1982). Similarly, it is suggested that the dorsal and ventral intermediate nuclei of reptiles are comparable to the VIA nuclei of birds and VA/VL nuclei of mammals (MEDINA et al., 1997).

The examples of connections cited above in marsupials, eutherian mammals and vertebrates outside the mammalian group (birds and reptiles) indicate that the ancestral neocortex of mammals had at least a convergence of somatosensory and motor thalamic projections to the anterior region of the parietal cortex. The concentration of cortico-spinal projections originating in this region indicates that the primary somaesthetic area could process motor information originating in the cerebellum and basal nuclei and command information to the dorsal aspect of the spinal cord via the cortico-spinal tract. These latter connections do not reach the motoneurons directly, but it is possible that they occur indirectly via the intemeurons located in the spinal cord (LEMON, 2008). The difficulty in finding a characteristic pattern of eutherian primary motor cortex in the amalgam region through electrophysiology, pattern of connections and histology can be explained by the fact that the species studied may have primitive characteristics combined with derived characteristics. We must remember that each species represents a "frozen moment" in the evolutionary process, with its own history and adaptations to its habitat. The neocortex of the opossum *Didelphis aurita* can be located at a stage from complete sensory-motor amalgamation to segregation of sensory-motor areas (Figure 42). Based on our results, we can say that the cortical region of the sensorimotor amalgam receives projections from thalamic nuclei that receive projections from the cerebellum and basal nuclei, although the location of its primary motor cortex is unclear so far.

55 Ebbessonian parcellation and the appearance of new cortical areas

The increase in the size of the neocortex over the course of mammalian evolution has triggered a series of changes in its organisation that contribute to many behavioural characteristics observed in mammals. When the neocortex increases in size in a species, its cortical areas will not necessarily increase in size on a larger scale (LARSEN and KRUBITZER, 2008). In fact, with encephalisation, cortical areas seem to undergo parcellation, and new areas can emerge in mammals (KAAS, 2004; LARSEN and KRUBITZER, 2008).

In larger brains, the number of neurons and the connections between them tend to increase (KAAS, 2004). As the distance between areas increases in these cases, the axons connecting different areas would need to be longer and thicker to maintain the same topography of connections and the same speed of information processing. A modification of this nature would generate a high energetic cost for the organism, as the brain would have to reserve more space for the axons. What actually seems to happen is that larger brains become more modular (with more areas), reducing the distance between neuron connections (KAAS, 2004, STRIEDTER, 2005).

On the other hand, in mammals with smaller brains, the total number of neurons tends to be smaller and if the brain maintained the same number of cortical areas, there would be fewer processing units (neurons) for each of them (KAAS, 2004). Reducing the number of non-primary areas seems to be an efficient mechanism for concentrating neurons in a single function (CATANIA et al., 1999; KAAS, 2004).

When we analyse the sensory-motor system of mammals, we see that ancestral mammals had a simple sensory-motor system and that many existing mammals have developed new and different characteristics over the course of evolution (KAAS, 2004, 2008).

The S1 and SC areas of the *Didelphis aurita* opossum in this study appear to be potential precursor regions for other sensorimotor areas, as they receive connections originating in the cerebellum and basal nuclei. In particular, the SC area also receives projections from the visual, auditory and somatosensory cortices. These connections to S1 or SC may have segregated as new areas emerged in the evolutionary process, generating the pattern of connections currently found in some mammals for areas M1, S1 and the posterior parietal cortex (REMPLE et al. 2007; PADBERG et al. 2005; PADBERG and KRUBITZER, 2006).

But what does the presence of the multimodal SC area in the opossum *Didelphis aurita* tell us? Did the cortical areas arise from a multimodal cortical organisation or did the multimodal area arise from a modality-specific cortical area?

Initially, it was believed that the neocortex was made up of a few primary and/or secondary areas and that these areas were surrounded by an associative, non-responsive cortex in anaesthetised preparations (KAAS, 1999). As larger brains had more non-responsive cortexes than smaller ones, it was assumed that the evolution of the brain was based on the addition of more associative fields to allow for more complex functions (KAAS, 2004). From the 1960s onwards, new studies indicated that the non-responsive regions in the mammalian neocortex, considered associative, actually performed unimodal sensory functions (HUBEL and WIESEL, 1965). Thus, in mammals with small brains, both multimodal areas and unimodal sensory and motor areas were identified, with the unimodal areas occupying the majority of the neocortex (KAAS, 2004). Since then, it has been believed that the evolution of complex brains occurred through the addition of high-hierarchical and unimodal sensory and motor areas (KAAS, 2004).

The possible mechanisms that trigger these changes in the neocortex remain under discussion. It is known that the cortical phenotype results from the interaction between various genetic and environmental factors (KRUBITZER and KAHN, 2003; KAAS and KRUBITZER, 2005; Krubitzer, 2007; LARSEN and KRUBITZER, 2008; SUR and RUBENSTEIN, 2005). The existence of a cortical organisation plan common to mammals in general reinforces the idea that the

formation of the primary cortical areas (and some secondary areas) would be genetically controlled (Figure 43) and that the possible modifications would be restricted (LARSEN and KRUBITZER, 2008; KRUBITZER and KAAS, 2005).

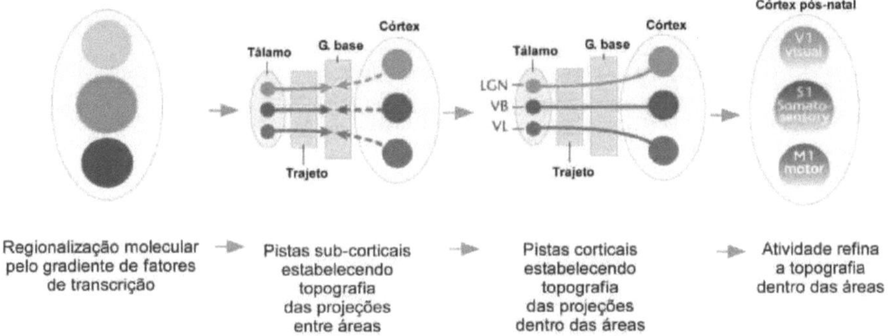

Figura 43. Development of cortical maps. During development, a gradient of transcription factors is created in the ventricular zone, generating an identity in the cortical plate and forming prototypes of cortical areas. The topography of the axons that grow from the cortical plate and the thalamus is regulated by the zones intermediate to the target (basal ganglion, for example). When these connections reach the target zone in the cortex, topographies are created within the areas, which are refined by mechanisms dependent on thalamic activity. Modified from Sur and Rubenstein, 2005.

For example, the localisation of cortical areas in the anteroposterior axis of the neocortex is controlled during development by the transcription factors Emx2 and Pax6 (SUR and RUBENSTEIN, 2005). Both transcription factors have their expression graded in the caudorostral (Emx2) and rostrocaudal (Pax6) axes (O'LEARY and NAKAGAWA, 2002). In some way, connections and other cortical characteristics are also managed by these genes. Emx2 and Pax6 regulate the specific expression of genes that code for cell adhesion molecules (Cad 6, 8 and 11), other transcription factors (Tbrl) and molecules (epinephrine-5) that guide axons to the cortex. It is also known that during development, these genes are involved in establishing the histological, functional, neuroanatomical and molecular identity of cortical areas. Even in the absence of thalamocortical activity, the activity of these genes persists, indicating an intrinsic regulation of these characteristics (LARSEN and KRUBITZER, 2008).

A classic example of gene regulation in cortical modules was demonstrated by the transposition of the signalling protein FGF-8 *in* the mouse *in utero* (FUKUCHI-SHIGOMORI and GROVE, 2001). This protein is localised in the rostral pole of the neocortex and was electroporated into a region caudal to SI, generating a doubling of barrel fields.

Following evolutionary reasoning, the duplication of barrels shown by Fukuchi-Shimogori and Grove (2001) may represent one of the mechanisms responsible for the subdivision of cortical areas (RAKIC, 2001). As we have seen, Ebbesson's parcellation hypothesis suggests that new cortical areas could arise during development by duplicating a precursor area. Consequently, the neural connections would reorganise, generating new patterns of projections characterising new areas (Figure 44) (STRIEDTER, 2005).

If the location of a cortical map is genetically determined during development and the trajectory of the thalamocortical projections is guided by molecular tracks located along the axon's path to the cortex (SUR and RUBENSTEIN, 2005) (Figure 43), the duplication of cortical areas could alter the distribution of the signalling molecules that attract the thalamic projections, which would generate a rearrangement of the projections to the cortex (STRIEDTER, 2005).

Analysing the characteristics of Ml in extant eutherian mammals, it is suggested that this area could have arisen from the duplication of a somatosensory area with thalamic connections also originating in the cerebellum and basal nuclei.

Our study was a pioneer in showing the connections of the SC area in the opossum *Didelphis aurita.* The similarity in the thalamocortical connections between Sl and SC, and the fact that both show responses to somesthetic stimulation, suggests that these areas may have a common cortical origin. Subsequent neuroanatomical studies of SC connections in other mammalian species will provide clues as to whether or not 1) this area was present in a supposed common ancestor of mammals, and whether it could have given rise to the mammalian sensory-motor system; or 2) it originated in the region of the sensory-motor amalgam.

Evolution of sensory-motor connections and areas

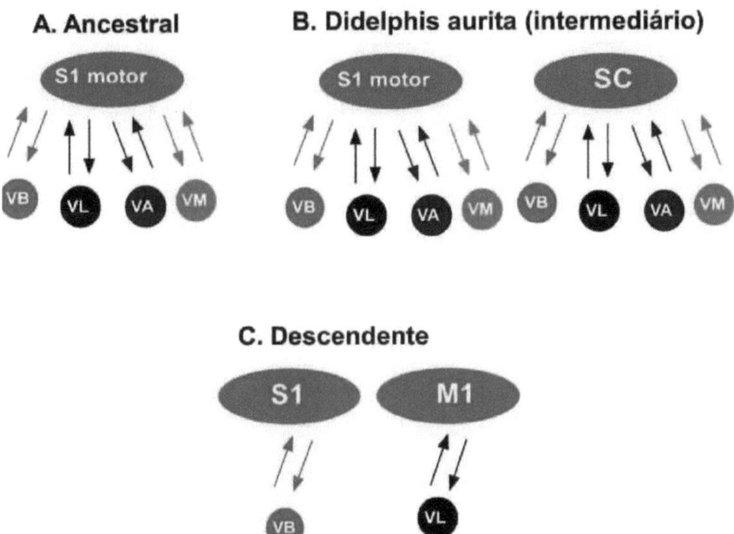

Figure 44: Ebbesson's parcel and the thalamocortical projections of the opossum *Didelphis aurita.* The ancestral mammalian encephalon could represent a sensory-motor amalgam condition, as seen in (A). With the subdivision of areas, the descendant species, with a new area similar to the precursor, would inherit the same pattern of connections (B) and could gradually develop a new pattern of thalamo-cortical or cortico-cortical projections. This would be the case of the opossum *Didelphis aurita,* whose thalamocortical projections from the ventral nuclei to Sl resemble the projections to SC, although the pattern is not completely identical. Similarly, the amalgam region could have given rise to areas Sl and Ml in extant eutherian mammals,

which would have developed their own thalamocortical connections throughout the evolutionary process (C).

114

CHAPTER 6

CONCLUSION

Before drawing any conclusions about the existence or not of a sensory-motor amalgam in the opossum *Didelphis aurita, we* need to understand what is being discussed here: the identification of a primary motor cortex as described for some eutherian and prototherian mammals, or a primary somesthetic region that processes motor information.

Undoubtedly, the parietal region immediately caudal to the orbital fissure has the characteristics of a somesthetic area and can be subdivided into SL, SR and SC. But we must consider that all anatomical and electrophysiological techniques must work together to confirm a primary motor area superimposed on the SL. As we have seen throughout this study, the electrophysiological data is inconclusive in this respect and leaves doubts about the presence of an exclusively motor somatotopic representation in the opossum.

It is possible that the primary motor cortex with the characteristics defined in today's eutherians arose during the evolution of mammals from a somatosensory cortex, such as the sensorimotor amalgam.

The primitive somesthetic cortex could process information that in eutherian mammals is carried out by areas of the motor system, without necessarily being named "motor cortex". This hypothesis is based on the observation of our data from the thalamocortical projections, the cortico-spinal projections and the histological data from Sl previously presented by other studies.

The sensory-motor cortical organisation of some species of non-mammalian animals receives projections from the thalamic nuclei, which receive projections from the cerebellar, basal and somatosensory nuclei without necessarily being segregated into somesthetic or primary motor areas. But we can go further...

In our work we found a somatosensory region that receives cortical projections from three different sensory modalities. However, little is known about whether these characteristics are present in all mammals or only in those whose brain is considered primitive. We can suggest that the appearance and specialisation of the motor and posterior parietal cortex throughout the evolutionary process has certainly contributed to the improvement of sensory and motor skills observed in mammalian behaviour.

BIBLIOGRAPHICAL REFERENCES

ALLMAN J.; ZUCKER S. **Cytochrome oxidase and functional coding in primate striate cortex: a hypothesis.** Cold Spring Harb Symp Quant Biol. 55: p. 979-82,1990.

ALLMAN, J.M.. **Evolving brains.** New York, Scientific American Library. 225.1999.

ANOMAL, R.F. **"Anatomo-functional organisation of the somesthetic cortex of the opossum *Didelphis aurita'*. Thalamocortical projections to areas Sl, SR and SC".** Master's thesis. 2005.

BECK, P.D.; POSPICHAL, M.W.; KAAS, J. H. **Topography, architecture and connections of somatosensory cortex in opossums: evidence for five somatosensory areas.** J. Comp. Neurol.366,p. 109-133,1996.

BODIAN, D. **Studies of the diencephalon of the *Virginia opossum.*** J. Comp. Neurol. 71, p. 259-323,1939.

BOHRINGER, R.C.; ROWE, M.J. **The organisation of the sensory and motor areas of the cerebral cortex in platypus (Ornithorrhyncus an a ti mus).** J. Comp. Neurol.174, p. 1-14,1977.

BRODMANN, K. **Vergleichande lolcaiisationslehre der grosshirnrinde.** Leipxing. East Germany: Baith. 1909.

BUDINGER E.; HEIL P.; SCHEICH H. **Functional organisation of auditory cortex in the Mongolian gerbil (Meriones unguiculatus). III. Anatomical subdivisions and corticocortical connections.** Eur J Neurosci. Jul;12(7): p. 2425-51,2000.

BUTLER, A.; HODOS, W. **Comparative Vertebrate Neuroanatomy.** Ed Wiley-liss. New York, 1996.

CARLSON, M.; WELT, C. **Somatic sensory cortex (Sml) of the prosimian primate Galago crassicaudatus: organisation of the mechanoreceptive input from hand in relation to architecture.** J. Comp. Neurol. 189, p. 249-271,1980.

CATANIA, K.C. **Cortical organisation in insectivore: The parallel evolution of the sensory periphery and the brain.** Brain Behav Evolution. 55: p. 311-321,2000.

CATANIA, K.C. **Evolution of sensory specialisations in insectivores.** The anatomical record. Part A, Discoveries in molecular, cellular, and evolutionary biology. 287(1): p. 1038-50,2005.

CATANIA, K.C.; JAIN, N.; FRANCA, J.G.; VOLCHAN, E.; KAAS. J.H. **The organisation of** somatosensory cortex in the short-tailed opossum *(Monodelphis domestica)*. Somatosensory and Motor Research, 171, p. 39-51,2000.

CATANIA K.C.; LYON D.C.; MOCK O.B.; KAAS J.H. **Cortical organisation in shrews: evidence from five species.** J Comp Neurol. 19;410(l): p. 55-72,1999.

CLARK, LE; GROS W.E. **The antecendents of man.** Edinburgh University Press, Edinburgh, UK. 1959.

COOKE, D.F.; TAYLOR, C.S.; MOORE, T.; GRAZIANO, M.S. **Complex movements evoked by microstimulation of the ventral intraparietal area.** Proc Natl Acad Sei. May. 13;100(10): p. 6163-8,2003.

COOPER, H.M.; HERBIN, M.; NEVO, E. Camp. Neural. 328, p. 313-350,1993.
CRAIG, A.D. and ZHANG, E.T. **Retrograde analyses of spinothalamic projections in the macaque monkey: input to posterolateral thalamus.** J Comp Neurol. 499: p. 953-964,2006.

CUSICK, C.G.; GOULD, H J. **Connections between area 3b of the somatosensory cortex and subdivisions of the ventroposterior nuclear complex and the anterior pulvinar nucleus in squirrel monkeys.** J Comp Neurol. Feb 1 ;292(1): p. 83-102,1990.

DARIAN-SMITH, C.; DARIAN-SMITH, L; CHEEMA, S.S. **Thalamic projections to sensorimotor cortex in the macaque monkey: use of multiple retrograde fluorescent tracers.** J Comp Neurol. Sep l;299(l):p. 17-46,1990.

DONOGHUE, J.P.; EBNER, F.F. **The organisation of thalamic projections of the parietal cortex of the Virginia opossum.** J.Comp. Neurol. 20, p. 365-388,1981.

EBBESSON S.O. **The parcellation theory and its relation to interspecific variability in brain organisation, evolutionary and ontogenetic development, and neuronal plasticity.** Cell and tissue research. 213(2): p. 179-212,1980.

ELDREDGE N.; CRACRAFT J. **Philogenetic patterns and evolutionary process.**
New York: Columbia University Press. 1980.

FABRI, M.; BURTON, H. **Ipsilateral cortical connections of primary somatic sensory cortex in rats.** J Comp Neurol. Sep 15;311(3): p. 405-24,1991.

FRANCA, J.G. **NADPH-diaphorase in the mammalian isocortex: Type I neurons and reactive neuroepithelia in the visual and somesthetic cortex.** Doctoral thesis. Health Sciences Centre - UFRJ, Rio de Janeiro, 1999.

FROST S.B.; MILLIKEN, G.W.; PLAUTZ, J.E.; MARSTERTON, B.; NUDO, R.

Somatosensory and motor representations in cerebral cortex of a primitive mammal *(Monodelphis domestica)'*, a window into the early evolution of sensorimotor cortex. **J.** Comp. Neurol. p. 421, 29-51,2000.
FUKUCHI-SHIMOGORI, T.; GROVE E.A. **Neocortex patterning by the secreted signalling molecule FGF8.** Science. Nov 2; 4(5544):p. 1071-4,2001.

FUNKE, K. **Somatosensory areas in the telencephalon of the pigeon. I. Response characteristics.** Exp Brain Res. 76(3): p. 603-19,1989.

GALLYAS, F. **Silver sataining of myelin by means of physical development.** Neurology. 1, p. 203-209,1979.

GRAY, P.A. **The Cortical lamination pattern of the opossum *Didelphis Virginiana.* J.** Comp. Neurol. 37, p. 221-263,1924.

HOOGLAND P.V. **Brainstem afferents to the thalamus in a lizard, Varanus exanthematicus.** The Journal of Comparative Neurology. 10; 210(2): p.152-62,1982.

HOROBIN, R. W. **Histochemistry.** Ed. Stuttgart, New York (1982).

HUBEL D.H.; WIESEL T.N. **Receptive fields and functional architecture in two nonstriate visual areas (18 and 19) of the cat.** J neurophysiol. 28:p. 229-89,1965.

HUFFMAN, K.J.; KRUBITZER, L. **Thalamo-cortical connections of areas 3a and Ml in Marmoset Monkeys.** J. Comp. Neurol. 435, p. 291-310,2001.

HUFFMAN, K.J. et al. **Organization of somatosensory cortex in three species of marsupials *Dasyurus hallucatus, Dactylopsila trivirgata* and *Monodelphis domestica'*, neural correlates of morphological specialisations.** J. Comp. Neurol. 403, p. 5-32,1999.

IVANCO, T.L.; PELLIS, S.M.; WHISHAW, I.Q. **Skilled forelimb movements in prey catching and in reaching by rats (Rattus norvegicus) and opossums (Monodelphis domestica): relations to anatomical differences in motor systems.** Behavi Brain Res. Sep, 79 (1-2): p. 163-81,1996.

JAIN N.; CATANIA K.C.; KAAS J.H. **A histologically visible representation of the fingers and palm in primate area 3b and its immutability following long-term deafferentations.** Cereb Cortex. 8: p. 227-236,1998.
JOHNSTON, J. B. **Further contributions of the study of evolution of the forebrain.** J. Comp. Neurol. 36: p.143-192,1923.

JONES, E.G. **The thalamus.** New York: Plenum Press. 1985.

JONES, E.G. **A new view of specific and nonspecific thalamocortical connections.** Advances in neurology. 1998;77: p.49-71,1998.

JONES, E.G.; WISE, S.P.; COULTER, J.D. **Differential thalamic relationships of sensory- motor and parietal cortical fields in monkeys.** The Journal of Comparative Neurology. 15;183(4): p.833-81,1979.

KAAS, J.H. **The segregation of function in the neurons Systems: Why do sensory Systems have so many subdivisions?** In: Contributions to sensory phisiology, W.P. Nef. New York Academic. 7, p. 201-240,1982.

KAAS, J.H. **What, if anything, is SI? Organisation of first somatosensory area of cortex.** Phisiologicals Rewiews. 63 (1) p. 206-231,1983.

KAAS, J.H. **The organisation of neocortex in mammals: implications for theories of brain function.** Annual review of psychology. 38: p. 129-51,1987.

KAAS, J.H. **How the somatosensory thalamus is subdivided and interconnected with areas of somatosensory cortex in monkeys?** Ed. M. Bentivoglio and R. Spreafico. Elsevier Science Publishers BV (Biomedical Division). p.143-50,1988.

KAAS, J.H. **Topographic maps are fundamental to sensory processing.** Brain Res Buli. 44(2): p.107-12,1997.

KAAS, J.H. **The transformation of association cortex into sensory cortex.** Brain Research Bulletin. 50(5-6): 425.1999.

KAAS, J.H. **Convergences in the modular and areal organisation of the forebrain of mammals: implications for the reconstruction of forebrain evolution.** Brain, Behavior and Evolution. 59, p. 262-272,2002.
KAAS, J.H. **Evolution of somatosensory and motor cortex in primates.** The anatomical record. Part A, Discoveries in molecular, cellular, and evolutionary biology. 281(1): p. 1148-56, 2004.

KAAS, J.H. **The evolution of the complex sensory and motor systems of the human brain.** Brain Research Bulletin. 18;75 (2-4): p. 384-90, 2008.

KAAS, J.H.; COLLINS, C.E. **Envisioning ideas of brain evolution.** Nature. 411, p. 141- 419,2001.

KAAS, J.H.; COLLINS, C.E. **The organisation of sensory cortex.** Current Opinion in Neuro biology. 11, p. 498-504,2001.

KAAS, J.H.; NELSON, RJ. **Connections of the ventroposterior nucleus of the thalamus with the body surface representations in cortical areas 3b and 1 of the cynomolgus macaque *(Macaca fascicularis)*.** J. Comp. Neurology. 199, p. 29-64,1981.

KAAS, J.H.; STEPNIEWSKA. **Encyclopedia of the human Brain.** Ed. Elsevier Science (USA), p.159-168,2002.

KAAS, J.H.; NELSON, RJ.; MERZENICK, M. **Multiple representations of the body within the primary somatosensory cortex of primates.** Science. 204; p. 521-523,1979.

KAAS, J.H.; NELSON, RJ.; MERZENICK, M. **Organisation of somatosensory cortex in primates. In: The organisation of the cerebral cortex,** F.O. Schimitt, F.G. Worden G. Adelman and S.G. Dennis. Boston, MA: MIT Press, p. 237-261,1981.

KALIL, K. **Neuroanatomical organisation of the primate motor system: afferent and efferent connections of the ventral thalamic nuclei.** Multidisciplinary Perspectives in the Event Related. D. Otton Washington. Brain Potential Research, p. 112-123,1978.

KARLEN, S.J.; KRUBITZER, L. **The functional and anatomical organisation of marsupial neocortex: evidence for parallel evolution across mammals.** Progress in Neurobiology.
82(3):p 122-41,2007.
KILLACKEY, H.; EBNER, F. **Convergent projections of the three thalamic nuclei onto a single cortical areas.** Science. 179, p. 283-285,1973.

KILLACKEY, H" RHOADES, R.W., BENNETT-CLARKE, C.A. **The formation of a cortical somatotopic map.** Trends in Neuroscience. 18(9): p. 402-7,1995.

KIMBLE, D.P. **Didelphid behaviour.** Neurosci Biobehav Rev. March;21(3): p. 361-9,1997. Review.

KORZENIEWSKA, E.; GÚNTÚRKÚN O. **Sensory properties and afferents of the N. dorsolateralis posterior thalami of the pigeon.** J Comp Neurol. 15;292(3): p. 457-79,1990.

KRUBITZER, L. **The organisation of neocortex in mammals: are species differences really so different?** Trends in Neuroscience. 18(9): p. 408-17,1995.

KRUBITZER, L. **The magnificent compromise: cortical field evolution in mammals.** Neuron. 25; 56(2): p. 201-8,2007.

KRUBITZER, L.A.; KAAS, J.H. **The organisation and connections of somatosensory cortex in marmosets.** J Neurosci. Mar;10(3): p. 952-74,1990.

KRUBITZER, L.A.; KAAS, J.H. **The somatosensory thalamus of monkeys: cortical connections and a redefinition of nuclei in marmosets.** J Comp Neurol. May 1 ;319(l):p. 123- 40,1992.

KRUBITZER,L.; KAAS, J. **The evolution of the neocortex in mammals: how is phenotypic diversity generated?** Current Opinion in Neurobiology. 15(4): p. 444-53, 2005.

KRUBITZER, L.; KAHN, D.M. **Nature versus nurture revisited: an old idea with a new twist.** Progress in Neurobiology. 70(1): p. 33-52, 2003.

KRUBITZER, L. KÚNZLE, H.; KAAS, J. **Organisation of sensory cortex in a Madagascan insectivore, the tenrec (Echinops telfairi).** J Comp Neurol. Mar 17; 379 (3), p. 399-414,1997.

KRUBITZER, L.; CALFORD,M.; SCHMID, L. **Connections of Somatosensory Cortex in Megachiropteran Bats: The Evolution of Cortical Fields in Mammals.** The Journal of Comparative Neurology. 327,p. 473-506 ,1993.

LARSEN, D.D.; KRUBITZER, L. **Genetic and epigenetic contributions to the cortical phenotype in mammals.** Brain Research Bulletin. 18;75(2-4): p. 391-7,2008.

LEMON, R.N. **Descending pathways in motor control.** Annu Rev Neurosci. 31: p. 195-218, 2008. Review.

LENDE, R.A. **Sensory representation in the cerebral cortex of the opossum *(Didelphis virginiana).*** J. Comp. Neurol. 121, p. 395-403,1963a.

LENDE, R.A. **Motor representation in the** cerebral cortex of the opossum *(Didelphis virginiana).* J. Comp. Neurol., 121, p. 405-415,1963b.

LENDE, R.A. **Representation in the cerebral cortex of the primitive mammal.** J. Neurophysiology. 27, p. 37-48, 1964.

LENDE, R.A. **A comparative approach to the neocortex: localisation in monotremes, marsupials and insectivores.** Ann NY Acad Sei 167: p. 262-275,1969.

LIVINGSTONE, M.S.; HUBEL, D.H. **Specificity of intrinsic connections in primate primary visual cortex.** Journal of Neuroscience. 4(11): p. 2830-5,1984.

M AG ALHÃES-C ASTRO, B.; SARA VIA, P.E. **Sensory and motor representation in the cerebral cortex of the marsupialis *Didelphis azarae azarae.*** Brain Research. 34, p. 291-299, 1971.

MARTIN, G.F.; FISHER, A.M. **A further evaluation of the origin, the course and the termination of the opossum corticospinal tract.** J Neurol Sei. Jul-Aug;7(l): p. 177-87,1968.
MARTINICH, S.; ROSA, M.G.; ROCHA-MIRANDA, C.E. **Patterns of cytochrome oxidase activity in**

the visual cortex of a South American opossum *(Didelphis marsupialis aurita).* Braz J Med Biol Res. 1990;23(9):p. 883-7,1990.

MARTINICH, S. **Morpho-functional organisation of the peristriate visual cortex of the opossum** *(Didelphis marsupialis).* Doctoral thesis. Health Sciences Centre - UFRJ, Rio de Janeiro, 1996.

MARTINICH, S.; PONTES, M.; ROCHA-MIRANDA, C.E. **Patterns of Corticocortical, Cortiçotectal, and Commissural Connections in the Opossum Visual Cortex.** The journal of comparative neurology. 416: p. 224-244,2000.

MEDINA, L.; VEENMAN, C.L.; REINER, A. **Evidence for a possible avian dorsal thalamic region comparable to the mammalian ventral anterior, ventral lateral, and oral ventroposterolateral nuclei.** J Comp Neurol. 21; 384(l):p. 86-108,1997.

MOUNTCASTLE, V. **Human physiology.** The C.V. Mosby Company. 1980.

MURPH, WJ.; EIZIRIK, E.; O' BRIEN, S.J.; MADESEN, O.; SCALLY, M.; DOUADY, C.J.; TEELING, E.; RYDER, A.O.; STANPHONE, M.J.; JONG, W.W.; SPRINGER, M.S. **Resolution of the early placental mammal radiation using Bayesian phylogenetics.** Science. 294, p. 2348-2351,2001.

NELSON, R.J.; KAAS J.H. **Connections of the ventroposterior nucleus of the thalamus with the body surface representation in cortical Areas 3b and 1 of the cynomolgus macaque** *(Macaca fascicularis).* J. Comp. Neurol. 199, p. 29-64,1981.

NORTHCUTT, R.G.; KAAS, J.H. **The emergence and evolution of mammalian neocortex.** Trends in Neuroscience. 18(9): p. 373-9,1995.

NUDO, R.J.; MASTERTON, R.B. **Descending pathways to the spinal cord, III: Sites of origin of the corticospinal tract.** J Comp Neurol. 22; 296(4):p. 559-83,1990.
OLIVEIRA, L. **Non-linear interpolation of contours in the primary visual cortex of the opossum** *(Didelphis aurita).* Thesis (Doctorate in Biological Sciences) - Health Sciences Centre - UFRJ, Rio de Janeiro, 2001.

O'LEARY, D.D.; NAKAGAWA Y. **Patterning centers, regulatory genes and extrinsic mechanisms controlling arealisation of the neocortex.** Current Opinion in Neurobiology. 12(l):p. 14-25,2002.

OSWALDO CRUZ, E.; ROCHA-MIRANDA, C.E. **The Brain of the opossum** *(Didelphis marsupialis).* Rio de Janeiro, Institute of Biophysics, Federal University of Rio de Janeiro, 1.1968.

OSWALDO CRUZ, E.; ROCHA-MIRANDA, C.E. **The diecephalon of the Opossum in Stereotaxic Coordinates. I. The epithalamus and dorsal thalamus.** J. Comp. Neurol. 129, p. 1-38,1968.

PEARSON, J.C.; HAINES, D.E. **Somatosensory thalamus of a prosimian primate (Galago senegalensis). II. An HRP and Golgi study of the ventral posterolateral nucleus (VPL).** J Comp Neurol. Apr 1;190(3):p. 559-80,1980.

PADBERG, J.; KRUBITZER, L. **Thalamocortical connections of anterior and posterior parietal cortical areas in New World titi monkeys.** The Journal of Comparative Neurology. 20;497(3):p. 416-35,2006.

PADBERG, J.; DISBROW, E.; KRUBITZER, L. **The organisation and connections of anterior and posterior parietal cortex in titi monkeys: do New World monkeys have an area 2?** Cerebral Cortex. 15(12): p. 1938-63,2005.

PALMER, A.R.; JIANG, D.; MCALPINE, D. **Neural responses in the inferior colliculus to binaural masking levei differences created by inverting the noise in one ear.** Joumal of Neurophysiology. 84(2): p. 844-52,2000.

PUBOLS, B.H.; PUBOLS, L.M.; DIPETTE, J.C.; SHEELY, J.C. **Opossum somatic sensory cortex: a microelectrode mapping study.** J. Comp. Neurol. 165, p. 229-246,1976.

RAKIC, P. **Neurocreationism-making new cortical maps.** Science. 2;294(5544): p. 1011-2, 2001. REES, S.; HORE, J. **The motor cortex of the brush-tailed possum (Trichosurus vulpecula): motor representation, motor function and the pyramidal tract.** Brain Res. Jun 15;20(3): p. 439-51,1970.

REINER, A.; VEENMAN, C.L.; MEDINA, L.;JIAO, Y.; DEL MAR, N.; HONIG, M.G. **Pathway tracing using biotinylated dextran amines.** Journal of Neuroscience Methods. 103, p. 23-37,2000.

REMPLE, M.S.; REED, J.L.; STEPNIEWSKA, L; KAAS, J.H. **Organisation of frontoparietal cortex in the tree shrew *(Tupaia belangeri)*. I. Architecture, microelectrode maps, and corticospinal connections.** The Journal of Comparative Neurology. 1 ;497(1): p. 133-54,2006.

REMPLE, M.S.; REED, J.L.; STEPNIEWSKA, L; LYON, D.C.; KAAS, J.H. **The organisation of frontoparietal cortex in the tree shrew (Tupaia belangeri): II. Connectional evidence for a frontal-posterior parietal network.** The Journal of Comparative Neurology. 1; 501(1): p. 121- 49,2007.

RIZZOLATTI, G.; FADIGA, L. **Grasping objects and grasping action meanings: the dual role of monkey rostroventral premotor cortex (area F5).** Novartis Found Symp. p. 218:81-95; discussion p. 95-103,1998.

ROCHA-REGO, V.; CANTERAS, N.S.; ANOMAL, R.F.; VOLCHAN, E.; FRANCA, J.G. **Architectonic subdivisions of the amygdalar complex of a primitive marsupial *(Didelphis aurita)*.** Brain Res Bull.l5;76(l-2): p. 26-35,2008.

ROSA, M.G.; KRUBITZER, L.A. **The evolution of visual cortex: where is V2?** Trends Neurosci. 22: p. 242-248,1999.

ROSA, M.G.; KRUBITZER, L.A.; MOLNÁR, Z.; NELSON, J.E.. **Organisation of visual cortex in the northern quoll, Dasyurus hallucatus: evidence for a homologue of the second visual area in marsupials.** The European journal of neuroscience. 11(3): p. 907-15,1999.

SCHERER-SINGLER, U.;VINCENT, S.R.; KIMURA, H.; MCGEER, E.G.Demonstration of a unique population of neurons with NADPH-diaphorase histochemistry. Joumal of Neuroscience Methods. 9(3): p. 229-34,1983.
SCHIEBER, M.H. **Constraints on somatotopic organisation in the primary motor cortex.** Joumal of Neurophysiology. 86(5): p. 2125-43,2001.

SCHROEDER, C.E.; LINDSLEY, R.W.; SPECHT, C.; MARCOVICI, A.; SMILEY, J.F.; JAVITT, D.C. **Somatosensory input to auditory association cortex in the macaque monkey.** J Neurophysiol. Mar;85(3): p. 1322-7,2001.

SLUTSKY, D.A.; MANGER, P.R.; KRUBITZER, L.A. **Multiple somatosensory areas in the anterior parietal cortex of the California ground squirrel (Spermophilus beechyii). J.** Comp. Neurol. 416, p. 521-539,2000.

SMITH, **E.G. The evolution of man.** Essays. Oxford University Press, London, UK. 1924.

STRIEDTER, G. **Principies of brain evolution.** Sinauer Associates, Inc. USA. 2005.

SOUSA, A.P.B.; OSWALDO CRUZ E.; R. GATTASS. **Somatotopic organisation and response properties of neurons of the ventrobasal complex in the opossum.** J. Comp. Neurol. 142, p. 231-248,1971.

STEPNIEWSKA, L; FANG, P.C.; KAAS, J.H. **Microstimulation reveals specialised subregions for different complex movements in posterior parietal cortex of prosimian galagos.** Proc Natl Acad Sei. Mar 29;102(13): p. 4878-83,2005.

STEPNIEWSKA, L; PREUSS, T.M.; KAAS, J.H. **Architetonic subdivisions of the motor thalamus of owl monkeys: Nissl, acetylcholinesterase, and cytochrome oxidase patterns. J.** Comp. Neurol. 349, p. 536-557,1993.

SUR, M.; RUBENSTEIN, J.L. **Patteming and plasticity of the cerebral cortex.** Science. 4;310(5749): p. 805-10, 2005.

THE AVIAN BRAIN NOMENCLATURE CONSORTIUMAvian brains and a new understanding of vertebrate brain evolution. Neuroscience. 6(2): p. 151-9,2005.

ULINSKY, P.S. **Thalamic projections to the somatosensory cortex of the echidna, (Tachyloglossus aculeatus).** In: Comparative neurology of the telencephalon. New York: Plenum Press, p. 449-481,1984.

VERCELLI, A.; REPICI, M.; GARBOSSA, D.; GRIMALDI, A. **Recent techniques for tracing pathways in the central nervous systems of developing and adult mammals.** Brain Research Bulletin. 50, p. 11-28,2000.

VOLCHAN, E.; FRANCA, J.G. **Distribution of NADPH-diaphorase-positive neurons in the opossum neocortex.** Brazilian journal of medical and biological research. 27(10): p. 2431-5, 1994.

WALLACE, M.N., RUTKOWSKI, R.G. and PALMER, A.R. **Identification and localisation of auditory areas in guinea pig cortex.** *Exp Brain Res* **132** (2000), p. 445^156,2000.

WALSH, T.M.; EBNER F.F. **Distribution of cerebellar and somatic lemniscal projections in the ventral nuclear complex of the Virginia opossum.** J Comp Neurol. Feb 15; 147(4): p. 427- 46,1973.

WATSON, C.R. *The corticospinal tract ofthe quokka wallaby (Setonix brachyurus).* J Anat. May;109(Pt 1): p. 127-33,1971.

WILEY, E.O. **Phylogenetics: the theory and practice of phylogenetic systematic...** John Wiley and Sons. New York. 1981.

WOOLSEY, C.N. **Organisation of somatic sensory and motor areas of the cerebral cortex.** In: Biological and Biochemical Bases of Behavior, H.F. Harlow and C.N. Woolsey. Madison, WI: Univ. of Wiconsin Press, p. 63-81,1958.

WOOLSEY, C.N.; FAIRMAN, D. **Contralateral, ipsilateral and bilateral representation of cutaneous receptors in somatic areas I and II of the cerebral cortex of pig, sheep and other mammals.** Surgery. 19, p. 684-702,1946.

WONG-RILEY, **M.Changes in the visual system of monocularly sutured or enucleated cats demonstrable with cytochrome oxidase histochemistry.** Brain Research. 27;171(l):p. 11-28, 1979.

Free Books

(httD://www.livrosaratis.com.br)

Thousands of books to download:

Download Management books
Download Agronomy books
Download Architecture books

Download Art books

Download Astronomy books

Download General Biology books

Download Computer Science books

Download Information Science books

Download Political Science books

Download Health Sciences books

Download Communication books

Download books by the National Education Council - CNE

Download civil defence books

Download law books

Download human rights books

Download economics books

Download Home Economics books

Download Education books

Download Education books - Traffic

Download Physical Education books

Download Aerospace Engineering books

Download Pharmacy books

Download Philosophy books

Download Physics books

Download Geoscience books

Download Geography books

Download history books

Download Language books

Download literature books

Download Cordel Literature books

Download Children's Literature books

Download maths books

Download medical books

Download Veterinary Medicine books

Download Environment books

Download Meteorology books

Download Monographs and Term Papers

Download Multidisciplinary books

Download music books

Download Psychology books

Download chemistry books

Download Collective Health books

Download Social Work books

I want morebooks!

Buy your books fast and straightforward online - at one of world's fastest growing online book stores! Environmentally sound due to Print-on-Demand technologies.

Buy your books online at
www.morebooks.shop

Kaufen Sie Ihre Bücher schnell und unkompliziert online – auf einer der am schnellsten wachsenden Buchhandelsplattformen weltweit! Dank Print-On-Demand umwelt- und ressourcenschonend produziert.

Bücher schneller online kaufen
www.morebooks.shop

info@omniscriptum.com
www.omniscriptum.com

Printed by Books on Demand GmbH, Norderstedt / Germany